108
Structure and Bonding

Series Editor:
D.M.P. Mingos

Springer-Verlag
Berlin Heidelberg GmbH

Supramolecular Assembly via Hydrogen Bonds I

Volume Editor:
D. M. P. Mingos

With contributions by
M. Alajarin · A.E. Aliev · A.D. Burrows · K.D.M. Harris
A. Pastor · J.W. Steed · D.R. Turner

Springer

The series *Structure and Bonding* publishes critical reviews on topics of research concerned with chemical structure and bonding. The scope of the series spans the entire Periodic Table. It focuses attention on new and developing areas of modern structural and theoretical chemistry such as nanostructures, molecular electronics, designed molecular solids, surfaces, metal clusters and supramolecular structures. Physical and spectroscopic techniques used to determine, examine and model structures fall within the purview of Structure and Bonding to the extent that the focus is on the scientific results obtained and not on specialist information concerning the techniques themselves. Issues associated with the development of bonding models and generalizations that illuminate the reactivity pathways and rates of chemical processes are also relevant.

As a rule, contributions are specially commissioned. The editors and publishers will, however, always be pleased to receive suggestions and supplementary information. Papers are accepted for *Structure and Bonding* in English.

In references *Structure and Bonding* is abbreviated *Struct Bond*
and is cited as a journal.

Springer WWW home page: http://www.springeronline.com
Visit the SB home page at http://www.springerlink.com/

ISSN 0081-5993
ISBN 978-3-642-05751-9 ISBN 978-3-540-39802-8 (eBook)
DOI 10.1007/978-3-540-39802-8

CIP Data applied for

springeronline.com

Originally published by Springer-Verlag Berlin Heidelberg New York in 2004
Softcover reprint of the hardcover 1st edition 2004

Typesetting: Fotosatz-Service Köhler GmbH, Würzburg
Production editor: Christiane Messerschmidt, Rheinau
Cover: Design & Production, Heidelberg

Printed on acid-free paper 02/3020 – 5 4 3 2 1 0

Series and Volume Editor

Editorial Board

Structure and Bonding is also Available Electronically

For all customers who have a subscription to Structure and Bonding, we offer the electronic version via SpringerLink free of charge. Please contact your librarian who can receive a password for free access to the full articles by registration at:

http://www.springerlink.com

If you do not have a subscription, you can still view the tables of contents of the volumes and the abstracts of each article by going to the SpringerLink Homepage, clicking on "Browse by Online Libraries", then "Chemical Sciences", and finally choose Structure and Bonding.

You will find information about the

- Editorial Board
- Aims and Scope
- Instructions for Authors
- Sample Contribution

at http://www.springeronline.com using the search function.

Preface

During the last two centuries synthetic chemists have developed a remarkable degree of control over molecular architecture. Currently organic and inorganic chemists are able introduce a wide range of substituents in predictable positions on increasingly more complex molecular scaffolds and even control the three dimensional stereochemistries at particular chiral centres. Indeed only the skill and imagination of an individual chemist limits the range of molecules he is able to produce. This process has been accelerated by the synergic nature of synthetic chemistry and spectroscopic and structural techniques which have confirmed the three dimensional structures of molecules.

A new frontier of chemistry has opened up in recent years which requires the development of analogous but new principles and methods which will enable chemists to predict how molecules interact with one another in the solid state. Indeed if we are to progress as "crystal engineers" as we have as "molecular engineers" we have to understand more predictively the factors which determine the three dimensional structures taken up by aggregates of molecules in the crystalline state. Therefore molecular recognition, material science, crystal engineering, nanotechnology, supramolecular chemistry the current goals of chemistry share the need to understand the very subtle factors which determine the way in which individual molecules come together in larger aggregates. In its most general form this is indeed a major problem because intermolecular forces are not very strong and are not very directional. However, this problem should be more amenable if there are groups on the surface of the molecules which are capable of hydrogen bonding. Not only are hydrogen bonds strong relative to other intermolecular forces but also they are more directional. Therefore, many groups have focussed their skills on the design of molecules with hydrogen bonding capabilities which can assemble in more predictable ways. These Volumes (108 and 111) bring together recent results from a range of leading research laboratories and define the current advances in this area. We still have a long way to go for a complete understanding, but these Volumes demonstrate that rapid and exciting progress is being made.

October 2003 D.M.P. Mingos

Contents

Probing Hydrogen Bonding in Solids Using Solid State NMR Spectroscopy
A.E. Aliev, K.D.M. Harris 1

Crystal Engineering Using Multiple Hydrogen Bonds
A.D. Burrows 55

Molecular Containers: Design Approaches and Applications
D.R. Turner, A. Pastor, M. Alajarin, J.W. Steed 97

Author Index 101–108 169

Subject Index 173

Contents of Volume 111

Supramolecular Assembly via Hydrogen Bonds II

Volume Editor: D.M.P. Mingos
ISBN 3-540-20086-x

Hydrogen Bonding Interactions Between Ions: A Powerful Tool in Molecular Crystal Engineering
D. Braga, L. Maini, M. Polito, F. Grepioni

Hydrogen-Bonded Supramolecular Chain and Sheet Formation by Coordinated Guanidine Derivatives
P. Hubberstey, U. Suksangpanya, M. Schröder

Hydrogen-Bonding Templated Assemblies
R. Vilar

Hydrogen Bonded Network Structures Constructed From Molecular Hosts
M.J. Hardie

Structure and Bonding, Vol. 108 (2004): 1–53
DOI 10.1007/b14136

Probing Hydrogen Bonding in Solids Using Solid State NMR Spectroscopy

Abil E. Aliev[1] · Kenneth D. M. Harris[2]

[1] Department of Chemistry, University College London, 20 Gordon Street, London WC1H 0AJ, United Kingdom
E-mail: A.E.Aliev@ucl.ac.uk

[2] School of Chemistry, University of Birmingham, Edgbaston Birmingham B15 2TT, United Kingdom
E-mail: K.D.M.Harris@bham.ac.uk

Abstract Solid state nuclear magnetic resonance (NMR) spectroscopy is a powerful and versatile technique for probing structural and dynamic properties of solid materials, and can provide detailed insights into the properties of hydrogen bonded systems. Of particular interest in this regard are solid state NMR experiments that investigate either the hydrogen atom directly involved in the hydrogen bond (employing ^{1}H NMR or ^{2}H NMR techniques) or the atoms within, or in close proximity of, the hydrogen bond donor and acceptor groups (employing, for example, ^{13}C, ^{15}N, ^{17}O, ^{29}Si or ^{31}P NMR techniques). To a large extent, the versatility of solid state NMR spectroscopy arises from this multinuclear capability, and the fact that there is considerable complementarity in the information that solid state NMR studies of different nuclei can provide. The aim of this chapter is to highlight some of the ways in which solid state NMR techniques, encompassing both traditional and recently developed methods, can be exploited towards understanding fundamental structural and dynamic properties of hydrogen bonded solids, focusing in particular on organic molecular materials.

Keywords Solid state NMR spectroscopy · Hydrogen bonding · Structure · Dynamics · NMR parameters

1 **Introduction** 3

2 **Solid State NMR Techniques for Studying Hydrogen Bonded Systems** 3

2.1 ^{1}H NMR 3
2.2 ^{2}H NMR 7
2.2.1 Lineshape Analysis 8
2.2.2 Spin-Lattice Relaxation 10
2.2.3 Two-Dimensional Exchange Spectroscopy 10
2.2.4 Selective Inversion 11
2.3 Dilute Spin $^{1}/_{2}$ Nuclei: ^{13}C, ^{15}N and ^{29}Si 11
2.4 ^{17}O NMR 13

3 **NMR Parameters and Hydrogen Bonding Geometry** 14

3.1 ^{2}H Quadrupole Coupling Constants 14
3.2 Isotropic ^{1}H Chemical Shifts 16
3.3 ^{1}H Chemical Shift Anisotropy 17
3.4 Isotropic ^{13}C Chemical Shifts and ^{13}C Chemical Shift Anisotropy . . 19
3.5 Isotropic ^{15}N Chemical Shifts and ^{15}N Chemical Shift Anisotropy . . 21

3.6 ^{17}O Chemical Shift and Electric Field Gradient Tensors 24
3.7 Overview 25

4 **Examples of Applications** 26

4.1 Structural Aspects of Hydrogen Bonding Arrangements in Solids . . 26
4.1.1 Carboxylic Acids 26
4.1.2 Peptides and Amides 27
4.1.3 Other Examples 31
4.2 Dynamic Aspects of Hydrogen Bonding Arrangements in Solids . . 33
4.2.1 Carboxylic Acids 33
4.2.2 Tropolone 37
4.2.3 Alcohols 39
4.2.4 Amino Acids, Peptides and Proteins 42
4.2.5 Urea, Thiourea and Their Inclusion Compounds 43
4.2.6 Pyrazoles, Imidazoles and Triazoles 44

5 **Concluding Remarks** 48

6 **References** 48

List of Abbreviations and Symbols

1D	One-dimensional
2D	Two-dimensional
CP	Cross polarisation
CRAMPS	Combined rotation and multiple pulse sequence
CSA	Chemical shift anisotropy
DANTE	Delays alternating with nutation for tailored excitation
DQ	Double quantum
EFG	Electric field gradient
EXSY	Exchange spectroscopy
IQNS	Incoherent quasielastic neutron scattering
IR	Infrared
MAS	Magic angle spinning
MQ	Multiple quantum
NMR	Nuclear magnetic resonance
NOE	Nuclear Overhauser effect
ODESSA	One-dimensional exchange spectroscopy by sideband alteration
PASS	Phase adjusted spinning sideband
REDOR	Rotational echo double resonance
SEDOR	Solid echo double resonance
δ_{iso}	Isotropic chemical shift
$\delta_{11}, \delta_{22}, \delta_{33}$	Chemical shift tensor components
Δ	Chemical shift anisotropy
Ω	Span of the chemical shift tensor

D	Dipolar coupling
e^2qQ/h	Quadrupole coupling constant
κ	Jump rate
T_1 ($T_{1\rho}$)	Spin-lattice relaxation time (in the rotating frame)

1 Introduction

Of the wide range of experimental methods that are utilised in the chemical sciences, nuclear magnetic resonance (NMR) spectroscopy is perhaps the most versatile, both in terms of the range of different types of systems and processes that can be studied and the wide variety of different types of information that can be obtained. This versatility is particularly exploited in applications of NMR spectroscopy to study solid materials. Given that each type of NMR active nucleus has an array of different properties, there is considerable complementarity in the information that solid state NMR studies of different nuclei can provide, including selective information on local structural properties and interactions and detailed information on different types of dynamic processes occurring across a broad range of characteristic timescales. Furthermore, studies of different types of NMR phenomenon or different types of NMR experiment for a given nucleus can again yield information on widely differing aspects of a material. With the continual development of new and increasingly ingenious solid state NMR techniques and pulse sequences, and the continued evolution of well established methodologies, there is considerable scope for applying solid state NMR to understand a very broad range of structural and dynamic aspects of solids.

Given the ubiquity of hydrogen bonding in chemical and biological systems and the many important phenomena that devolve upon hydrogen bonding, it is not surprising that NMR techniques have been used widely to understand structural and dynamic aspects of hydrogen bonded systems, and the aim of this chapter is to give an overview of some of the ways in which solid state NMR spectroscopy can be employed in this regard. We focus primarily, although not exclusively, on hydrogen bonding within organic molecular crystals, and we place emphasis on applications of techniques and the types of information that they can reveal, rather than on fundamental aspects of the techniques themselves. Details of NMR phenomena in general [1–6] and solid state NMR techniques in particular [7–13] can be found in the cited references.

2 Solid State NMR Techniques for Studying Hydrogen Bonded Systems

2.1 ^{1}H NMR

In principle, ^{1}H NMR might be expected to be the most appropriate technique for studying both structural and dynamic aspects of hydrogen bonding in solids, by directly probing the hydrogen atoms involved in hydrogen bonds. However, it is generally difficult to record high-resolution ^{1}H NMR spectra of solids as the very strong homonuclear ^{1}H-^{1}H dipole-dipole interaction leads to spectra that are typically broad and featureless. The homonuclear ^{1}H-^{1}H dipolar interaction is usually of the order of 50 kHz and is the dominant anisotropic interaction governing both the ^{1}H NMR lineshape and ^{1}H relaxation. The magnitude of this interaction depends directly on the ^{1}H$\cdots$^{1}H internuclear distance, and can therefore be used to derive information on distances between ^{1}H nuclei in a solid. However, in most organic solids, there are many different ^{1}H nuclei in close proximity of each other, and the multitude of different ^{1}H-^{1}H dipole-dipole interactions gives rise to severe broadening of the spectrum. If the ^{1}H NMR spectrum is dominated by a single ^{1}H-^{1}H dipole-dipole interaction (for example, by use of appropriate selectively deuterated materials), analysis of the ^{1}H NMR spectrum becomes straightforward and the following expression can be used to derive the internuclear distance of interest:

$$\Delta\nu_{dip} = \frac{3}{2}\frac{h\gamma^2}{(2\pi)^2 r_{\mathrm{HH}}^3}\frac{\mu_0}{4\pi}(3\cos^2\theta - 1) \qquad (1)$$

where r_{HH} is the ^{1}H-^{1}H internuclear distance, θ is the angle specifying the orientation of the ^{1}H-^{1}H internuclear vector with respect to the magnetic field direction, γ is the magnetogyric ratio for ^{1}H, h is Planck's constant and μ_o is the permeability constant ($4\pi\times10^{-7}$ kg m s^{-2} A^{-2}).

When the ^{1}H-^{1}H dipole-dipole interaction can be measured for a specific pair of ^{1}H nuclei, studies of the temperature dependence of both the ^{1}H NMR lineshape and the ^{1}H NMR relaxation provide a powerful way of probing the molecular dynamics, even in very low temperature regimes at which the dynamics often exhibit quantum tunnelling behaviour. In such cases, ^{1}H NMR can be superior to quasielastic neutron scattering experiments in terms of both practicality and resolution. The experimental analysis can be made even more informative by carrying out ^{1}H NMR measurements on single crystal samples. In principle, studies of both the ^{1}H NMR lineshape and relaxation properties can be used to derive correlation times (τ_c) for the motion; in practice, however, spin-lattice relaxation time (T_1) measurements are more often used to measure τ_c as they are sensitive to the effects of motion over considerably wider temperature ranges.

As an example, we consider ^{1}H NMR measurements on a single crystal of benzoic acid [14], carried out to investigate tunnelling dynamics in hydrogen bonded carboxylic acid dimers.

For the partially deuterated benzoic acid (C_6D_5COOH), the solid state ^{1}H NMR spectrum is dominated by the intra-dimer ^{1}H-^{1}H dipole-dipole interaction. In a single crystal, both tautomers A and B are characterised by a well-defined interproton vector with respect to the direction of the magnetic field (Fig. 1). Proton motion modulates the ^{1}H-^{1}H dipole-dipole interactions, which in turn affects the ^{1}H NMR lineshape and the spin-lattice relaxation time. It has been shown that spin-lattice relaxation times are sensitive to the proton dynamics over the temperature range from 10 K to 300 K, and at low temperatures incoherent quantum tunnelling characterises the proton dynamics. A dipolar splitting of about 16 kHz is observed at 20 K. From the orientation dependence of the dipolar splitting, the

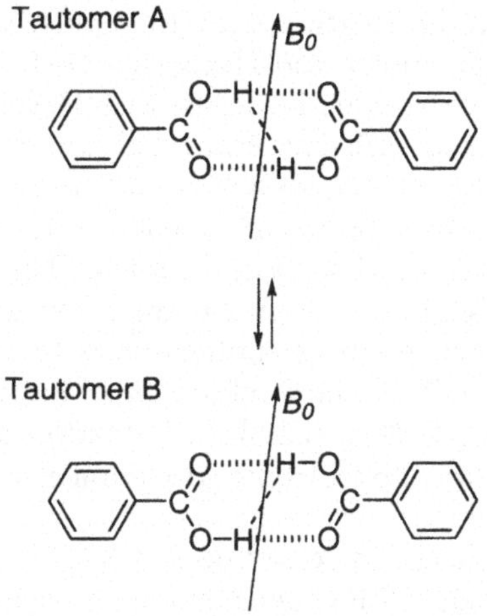

Fig. 1 The two tautomers of the benzoic acid dimer. For each species the orientation of the vector connecting the two protons relative to the direction of the applied magnetic field (B_0) is different [14]

$^1H\cdots{}^1H$ distance in the dominant tautomer (A) at this temperature was established to be 2.26±0.08 Å, in good agreement with neutron diffraction results [15]. A subsequent paper [16] considered spin-lattice relaxation theories for classical and tunnelling motions of protons and showed that only one of these theories provides a satisfactory explanation for the experimentally determined frequency dependence of the ^1H spin-lattice relaxation rates for benzoic acid and a few benzoic acid derivatives.

The above example illustrates how wideline ^1H NMR can be used to investigate aspects of both the structure and dynamics of hydrogen bonded solids. In this case, resolution capacity of the technique was provided by the sample itself, by use of a partially deuterated single crystal sample. The main advantage of ^1H NMR for a static solid sample is that the measurements can be extended to extremely low or high temperatures. This is important in many cases, as very often an evolution of hydrogen bonding dynamics over a large temperature range and interconversion between different forms of dynamics is of interest.

A related approach for recording high-resolution ^1H NMR spectra of solids is to use a fully deuterated sample. In this case, the residual impurity ^1H nuclei are detected, but because these nuclei are spatially dilute, homonuclear ^1H-^1H dipole-dipole interactions (which fall off rapidly with internuclear distance) are weak, and hence narrow lines can be obtained in ^1H NMR spectra of such materials.

However, preparation of partially or fully deuterated materials may not always be feasible and single crystal samples are not always available. Hence, techniques have been developed that allow high-resolution solid state ^1H NMR spectra to be recorded for powder samples without any isotope enrichment. Homonuclear ^1H-^1H dipole-dipole interactions are typically of the order of 50 kHz leading to very broad spectra as discussed above. In order to alleviate the effects of the homonuclear dipole-dipole interactions and obtain a ^1H NMR spectrum that conveys ^{1}H chemical shift information, a multiple pulse sequence approach was first developed by Waugh, Huber and Haeberlen [17]. The multiple pulse sequence developed by these authors is known as WHH-4 and has been used in subsequent developments for efficient removal of homonuclear dipolar interactions. These multiple pulse techniques achieve line narrowing by manipulation of the spin operators for the appropriate nuclear interactions, rather than by modification of the spatial coordinates of the nuclei. The resolution in these experiments can be further enhanced by employing conventional magic angle sample spinning (MAS), which removes contributions to line broadening due to chemical shift anisotropy (CSA). The combination of the multiple pulse sequence with MAS is known as CRAMPS (combined rotation and multiple pulse sequence) technique [18–20], and isotropic ^{1}H chemical shifts can be resolved in the resulting spectrum.

Despite numerous applications, conventional CRAMPS still remains one of the most demanding solid state NMR experiments as it requires the use of specially prepared spherical samples to minimise radiofrequency inhomogeneity effects and the careful calibration and setting of pulse widths and phases. Further modifications of the experiment that do not require the complicated and extended set-up procedures have been suggested recently. These are known as rotor-synchronised CRAMPS, which combines a new multiple pulse sequence [21], and its modification which uses a standard WHH-4 sequence at ultrafast MAS frequencies (e.g. 35 kHz) [22].

An advantage of the new rotor-synchronised CRAMPS experiment is that isotropic ^{1}H chemical shifts (which can convey considerable information in relation to hydrogen bonding) can be derived directly from the spectrum. However, the homonuclear dipolar coupling between protons, which can be used to assess the through-space proximity of protons (and is therefore of potential interest in the study of hydrogen bonded systems), is suppressed in these experiments. An experiment that has the resolution capacity of CRAMPS but also allows measurement of dipolar interaction strengths for different proton pairs would therefore be highly desirable. Recently developed two-dimensional (2D) ^{1}H double quantum (DQ) MAS NMR [13] largely fulfils this requirement, although the level of information available from these spectra strongly depends on the resolution in both frequency dimensions. Nevertheless, for hydrogen bonded solids, protons involved in hydrogen bonding resonate at considerably higher frequencies, hence 2D DQ MAS techniques are suitable in the majority of cases. In terms of practical implementation, the concept of suppression of dipolar couplings used in the CRAMPS experiment is applied in reverse in the DQ MAS experiments with the aim of reintroducing the necessary dipolar coupling during the excitation and reconversion periods of the experiment. The interpretation of the resulting 2D DQ MAS spectrum is straightforward and is similar to that of solution state 2D NMR spectra: the DQ frequency corresponding to a given DQ coherence is simply the sum of the two single quantum frequencies (i.e. chemical shifts) and the presence of a DQ peak implies a close spatial proximity of the two ^{1}H nuclei involved.

The efficiency of DQ excitation in the DQ MAS experiments is proportional to the square of the dipolar coupling. The dipolar coupling D is proportional to

$(r_{HH})^{-3}$, hence the integrated intensity of the DQ peak is inversely proportional to the sixth power of the internuclear distance [I_{DQ}~$(r_{HH})^{-6}$], a relationship similar to that widely used in solution state NMR [NOE~$(r_{HH})^{-6}$]. As a result, proton pairs with internuclear distances in the range 1.8–3.5 Å can be studied using this technique. As reported recently, highly accurate measurements of ^{1}H-^{1}H internuclear distances (to within ±0.02 Å) can be achieved by the 2D DQ MAS technique [23]. Furthermore, torsion angles can also be determined using this method [24].

Note that the ability of the ^{1}H DQ MAS technique to determine accurately the ^{1}H-^{1}H distances in hydrogen bonded solids can also provide an independent distance constraint for the refinement of the structure from X-ray diffraction data [25]. This is somewhat analogous to the use of NOE constraints measured by solution NMR for protein structure determination and promises to be a widely applicable approach once techniques based on ^{1}H DQ MAS are routinely available.

2.2
^{2}H NMR

In general, the replacement of protons by deuterons has negligible effect on the structure and dynamics of a solid material, but given that the NMR properties of ^{1}H and ^{2}H nuclei differ substantially, ^{1}H and ^{2}H NMR spectroscopic techniques are complimentary to one another, and each has specific advantages for investigating different types of systems or processes. For example, ^{2}H is a quadrupolar nucleus (spin I=1), and ^{2}H NMR spectra are generally dominated by the quadrupolar interaction that occurs between the nuclear quadrupole moment and the electric field gradient (EFG) at the nucleus. On the other hand, the ^{1}H nucleus (spin I=1/2) is not quadrupolar. Although an obvious shortcoming of ^{2}H NMR is the need to synthesise ^{2}H enriched materials, for molecules containing functional groups of reasonable acidity (e.g. hydroxyl or amino groups), selective deuteration of these functional groups can be readily carried out.

^{2}H NMR is a powerful technique for studying both structural and dynamic properties of hydrogen bonded solids. As discussed below, the ^{2}H quadrupole coupling constant was one of the first NMR parameters for which convincing correlations were found with hydrogen bond geometry. A new experimental approach for highly precise measurements of ^{2}H quadrupole interaction parameters, as well as the ^{2}H chemical shift tensor, has been reported recently [26], and illustrated for deuterated calcium formate, α-Ca(DCOO)$_2$.

Although solid state ^{2}H NMR techniques are also used widely in structural studies, the principal use of these techniques has been to obtain detailed information on reorientational motions in the solid state, and our discussion is focused on this aspect of ^{2}H NMR. As discussed above, the quadrupole interaction is usually the dominant nuclear spin interaction in ^{2}H NMR, and other nuclear spin interactions (e.g. dipole-dipole interaction, CSA and scalar J-coupling) are generally negligible in comparison. For ^{2}H, the quadrupole interaction is typically about 150–250 kHz, whereas the direct dipolar interactions and CSA are typically about 10 kHz and 0.7 kHz (at 11.7 T) respectively. Since the EFG originates from the distribution of charges in the molecule and its close vicinity in the solid, it is primarily intramolecular in nature, and ^{2}H NMR spectra are particularly sensitive to molecular reorientational processes in solids (in particular, reorientation of the bond containing the ^{2}H nucleus). The dynamic range over which motional effects can be followed is extremely large in the case of ^{2}H NMR due to the availability of various complementary ^{2}H NMR techniques. For example, dynamic studies can be carried out using ^{2}H NMR spin-alignment techniques (for motions with frequencies in the range 10^{-2}–10^3 Hz), lineshape analysis (for motions with frequencies in the range 10^3–10^8 Hz) and spin-lattice relaxation time measurements (for motions with frequencies in the range 10^6–10^{11} Hz). As a consequence, ^{2}H NMR has become one of the most widely applied techniques for studying the dynamics of hydrogen bonded systems.

We now consider some specific features of the most widely used ^{2}H NMR technique – lineshape analysis – as well as other important ^{2}H NMR techniques. More detailed discussion can be found in other review articles [27, 28]. By employing appropriate combinations of these techniques, and exploiting the complementarity between them, a detailed understanding of the dynamic properties may be established.

2.2.1
Lineshape Analysis

^{2}H NMR lineshape analysis is probably the most widely applied technique for studying dynamic properties of organic solids. The basis of this approach is that, when a ^{2}H nucleus undergoes motion on an appropriate timescale, the ^{2}H NMR lineshape is altered in a well-defined manner, allowing detailed mechanistic information on the dynamic process to be elucidated. When the rate of molecular motion is intermediate on the ^{2}H NMR timescale (i.e. frequency of motion between ca. 10^3 and 10^8 Hz), the appearance of the ^{2}H NMR spectrum depends critically upon the exact rate and geometry of the molecular reorientational motion. The dependence of the ^{2}H NMR spectrum on the dynamic properties of the ^{2}H nucleus can be simulated theoretically [29, 30], and the purpose of lineshape analysis of a set of experimental ^{2}H NMR spectra, recorded as a function of temperature, is to propose plausible mechanisms for the motion and then for each of these mechanisms: (i) to simulate theoretically a set of ^{2}H NMR spectra corresponding to different values of the rate of motion, and (ii) to decide whether the set of simulated ^{2}H NMR spectra is in satisfactory agreement with the set of experimental spectra. If the rate of molecular motion is intermediate on the ^{2}H NMR timescale, then it is possible to determine the rate of motion as an accurate function of temperature (by finding, in stage (ii), the simulated spectrum that best fits the experimental spectrum recorded at each temperature). A hurdle in this approach is finding the best fits to the experimental spectra and so far the approach adopted has been based on trial and error variation of parameters coupled with fitting "by eye" (i.e. visual comparison to assess the quality of fit between the simulated and the experimental spectra). However, complex cases of lineshape analysis may require variation of several parameters in spectral simulations. A more efficient, automated approach for lineshape fitting has been

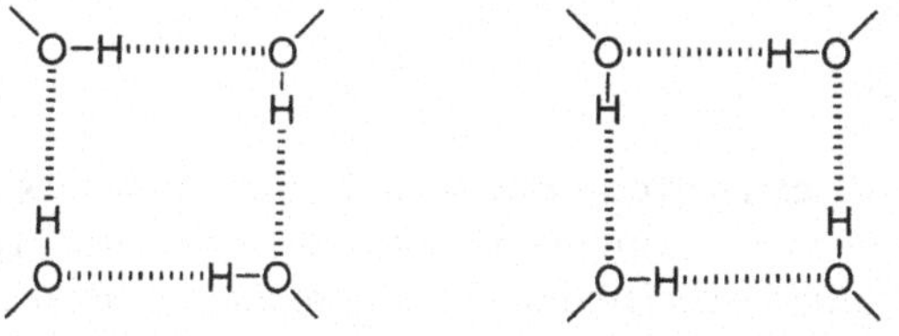

Fig. 2 Schematic representation of the hydrogen bonding arrangement in the dimer of ferrocene-1,1′-diylbis(diphenylmethanol) [33]

reported recently [31]. This approach provides an objective assessment of the level of agreement between experimental and simulated ^{2}H NMR spectra, and removes much of the subjectivity that is characteristic of the traditional approach of comparing experimental and simulated ^{2}H NMR spectra "by eye".

The widespread use of ^{2}H NMR lineshape analysis has also revealed certain limitations of the technique. Care is needed in interpreting lineshape changes, particularly for systems with complicated motional processes. In some cases, even though a proposed dynamic model may give a good fit to the experimental spectra, other plausible dynamic models might also be able to fit the experimental spectra [32]. Establishing an unambiguous and unique assignment of the dynamic process is therefore an important issue, and in many cases results from other experimental techniques (including other NMR techniques) must be used to distinguish between postulated models. However, we note that the range of timescales (10^{-3}–10^{-8} s) covered by ^{2}H NMR lineshape analysis is not within easy reach of other widely used physical techniques such as neutron, Raman and Brillouin scattering, as well as molecular dynamics (MD) simulations.

An example concerns the hydrogen bond dynamics in selectively deuterated ferrocene-1,1′-diylbis(diphenylmethanol-d_1). In this structure, the molecules form hydrogen bonded dimers, with the oxygen atoms of four hydroxyl groups involved in a folded trapezium hydrogen bonding arrangement [33] shown schematically in Fig. 2 as a square.

Each hydroxyl hydrogen atom is disordered between two equally populated positions, from which it is inferred that there are two plausible arrangements ("clockwise" and "anticlockwise") of the hydrogen bonded unit. From ^{2}H NMR lineshape analysis and ^{2}H NMR spin-lattice relaxation time measurements, the dynamic properties of the hydroxyl deuterons are equally consistent with the following dynamic models: (i) hydrogen transfer between adjacent hydroxyl oxygen atoms, and (ii) a 2-site 180° jump motion of each hydroxyl group about its C-O bond. In general, these dynamic models can be distinguished on the basis of ^{2}H NMR, but for the specific geometry of the intermolecular hydrogen bonding arrangement in this solid, these models fit the ^{2}H NMR data equally well.

2.2.2 *Spin-Lattice Relaxation*

In early years of NMR, extensive studies of molecular dynamics were carried out using relaxation time measurements for spin 1/2 nuclei (mainly for ^{1}H, ^{13}C and ^{31}P). However, difficulties associated with assignment of dipolar mechanisms and proper analysis of many-body dipole-dipole interactions for spin 1/2 nuclei have restricted their widespread application. Relaxation behaviour in the case of nuclei with spin greater than 1/2 on the other hand is mainly determined by the quadrupolar interaction and since the quadrupolar interaction is effectively a single nucleus property, few structural assumptions are required to analyse the relaxation behaviour.

The use of ^{2}H NMR spin-lattice relaxation time measurements allows dynamic processes with motional frequencies between $\nu/10^3$ and $\nu/10^{-3}$ to be studied [2], where ν denotes the Larmor frequency of the ^{2}H nucleus. Thus, molecular motions in the frequency range 10^4–10^{11} Hz are typically studied using ^{2}H spin-lattice relaxation time measurements. Note that in many cases ^{2}H NMR lineshapes characteristic of the rapid motion regime (with motional frequencies greater than 10^8 Hz) are observed at temperatures as low as 77 K (which is the lowest temperature attainable on solid state NMR spectrometers equipped with liquid nitrogen cryostats). In such cases, lineshape analysis techniques [which are particularly informative for establishing details of dynamic processes in the intermediate motion regime (motional frequencies 10^3–10^8 Hz)] can only provide limited information on the dynamic properties, leaving spin-lattice relaxation time measurements as the only choice.

Theoretical expressions for spin-lattice relaxation of ^{2}H nuclei (determined by locally axially symmetric quadrupolar interactions modulated by molecular motions) can be derived for specific dynamic processes, allowing the correct dynamic model to be established by comparison of theoretical and experimental results [34, 35]. In addition, T_1 anisotropy effects, which can be revealed using a modified inversion recovery experiment, can also be informative with regard to establishing the dynamic model [34, 35].

2.2.3 *Two-Dimensional Exchange Spectroscopy*

The motional timescales that are accessible by solid state ^{2}H NMR are further extended by 2D exchange techniques, which permit the investigation of ultraslow motions occurring at frequencies of the order of 10^3–10^{-2} Hz [36, 37]. This presents additional possibilities for detailed investigation of dynamic processes that are in the "static" motional regime with respect to the conventional ^{2}H NMR technique discussed above. Other advantages of these 2D solid state ^{2}H NMR experiments are: (i) geometrical information (e.g. jump angles) describing the motion of the ^{2}H nucleus can be determined directly from the spectrum, and (ii) jump motions and diffusive motions can be distinguished directly.

The performance has been further developed [38] leading to improved pulse sequences for multidimensional solid state exchange NMR. In particular, the use of five-pulse sequences greatly facilitates processing of the spectra and decreases phase distortions and artefacts in the spectra. Finally, a new 1D NMR exchange experiment (which consists of the usual three-pulse sequence for 2D exchange spectroscopy) in the slow motion regime of spinning solids, with chemically equivalent nuclei exhibiting quadrupole coupling, has been proposed [39].

2.2.4
Selective Inversion

Frequency selectivity in wideline ^{2}H NMR studies can be achieved using selective inversion of a narrow band of frequencies using either a DANTE (delays alternating with nutation for tailored excitation) sequence of hard pulses with small flip angle or a weak inversion pulse with long pulse length [40]. Selective inversion allows selection of frequency domains of the ^{2}H NMR powder pattern corresponding to specific orientations of the principal axis of the EFG tensor with respect to the external field direction. In the presence of molecular motions, the selectively inverted spins can jump to orientations outside the excited frequency range and this can provide information about the jump angle. In principle, selective inversion can provide the same type of information about the geometry of motion as the 2D exchange technique discussed above, although model-dependent lineshape simulations are required in the case of the selective inversion technique. On the other hand, the rate of slow molecular motions can be easily and accurately measured using selective inversion techniques. In addition, the optimal mixing time for obtaining a 2D exchange spectrum can be determined quickly using selective inversion.

Aspects of the experimental implementation of the technique have been studied in detail [41] and it has been shown that double sideband modulation and pulse shaping can be combined to improve the performance of selective pulses in solid state ^{2}H NMR. Applications of the selective inversion-recovery experiment using a DANTE sequence to study ultraslow motions have been demonstrated [42, 43].

2.3
Dilute Spin $^{1}/_{2}$ Nuclei: ^{13}C, ^{15}N and ^{29}Si

Traditional NMR techniques for spin 1/2 nuclei, such as ^{13}C, ^{15}N and ^{29}Si, are well known and are routinely applied. In such cases, high-resolution solid state NMR spectra are generally recorded *via* a combination of MAS and high-power ^{1}H decoupling. In appropriate cases, cross polarisation (CP) from an abundant spin such as ^{1}H is also employed to enhance the sensitivity of the technique. With these techniques, narrow resonance lines are obtained in the NMR spectrum. In principle, the high-resolution solid state NMR spectrum will contain one peak for each crystallographically distinguishable environment of the nucleus under investigation (^{13}C, ^{15}N, ^{29}Si or ^{31}P) in the crystal structure. However, if MAS is not sufficiently rapid, the NMR signal for a given ^{13}C environment comprises a peak at the isotropic chemical shift, and a series of "spinning sidebands" displaced from the "isotropic peak" by integer multiples of the MAS frequency. The position of the isotropic peak is independent of the MAS frequency, whereas the positions of the spinning sidebands vary as the MAS frequency is varied. Intensities of the spinning sidebands depend on the ^{13}C CSA and using Herzfeld-Berger analysis, the chemical shift tensor components can be estimated from the spinning sideband manifold. Several aspects of high-resolution solid state ^{13}C NMR can be exploited to investigate dynamic properties, including: (i) lineshape analysis, (ii) 2D exchange techniques and (iii) relaxation time measurements. These and other routine applications of solid state NMR techniques have been covered in a recent review [12].

Advanced techniques of specific utility in the case of hydrogen bonded solids have been reported, and are now highlighted. A new approach based on 2D ^{1}H-^{13}C heteronuclear correlation spectroscopy with a CP sequence [44, 45] has been used to study C=O···H-N and C(O)-O-H···O=C hydrogen bonding interactions in amino acids and peptides [46]. It has been shown that the cross-peak volumes in the 2D spectra correlate with the C···H distance and can be used to estimate distances with a standard deviation of 0.2 Å. The upper limit for the distance estimation is 3 Å, which is sufficiently large to cover the range of hydrogen bonding distances. Additionally, ^{1}H and ^{13}C chemical shift information can be derived from these spectra, both of which are sensitive to hydrogen bonding effects.

Interesting information about hydrogen bonded structures has been obtained by NMR experiments that utilise cross polarisation from ^{1}H to ^{29}Si, allowing hydrogen bonding of silanols on silica surfaces to be studied by ^{1}H-^{29}Si CP MAS NMR [47], in which cross polarisation efficiency was used to estimate heteronuclear dipolar interaction strengths. The critical parameter in the CP studies is the ^{1}H-^{29}Si cross polarisation rate constant $(T_{HSi})^{-1}$ which is easily measured from experiments carried out as a function of the CP contact time. This rate constant depends on the strengths of the ^{1}H-^{29}Si and ^{1}H-^{1}H dipolar interactions, and is roughly proportional to the inverse sixth power of the ^{1}H-^{29}Si internuclear distance. It has been shown that for the ^{29}Si nuclei of isolated single silanols, the CP time constant T_{HSi} is at least five times larger than that for hydrogen bonded silanols [48]. The single silanols with T_{HSi}=14 ms were assigned as not hydrogen bonded, whereas those with T_{HSi}=1.2 ms were assigned as hydrogen bonded single silanols. Similarly, T_{HSi} values of 6 ms and 0.5 ms were assigned to non-hydrogen bonded and hydrogen bonded geminal silanols, respectively. Clearly, this methodology is also applicable for other commonly studied "dilute" nuclei such as ^{13}C and ^{15}N, especially when there are no protons directly bonded to the nucleus of interest.

Highly accurate interatomic distances (ultimately ±0.05 Å) may be obtained from REDOR experiments [49], which are therefore an attractive tool for studies of hydrogen bonding. This technique has been used recently to characterise α-helix structures in polypeptides by measuring ^{13}C=O···H-^{15}N hydrogen bond lengths [50]. The intrachain ^{13}C···^{15}N interatomic distances, measured for a number of different samples, were found to be 4.5±0.1 Å. This finding was used as evidence for the α-helix structure, which is consistent with the conformation dependent displacements of ^{13}C chemical shifts of the C_{α}, C_{β} and carbonyl carbons of the peptide unit [51].

A comparatively long N···H hydrogen bond length in the benzoxazine dimer (see below), measured using an advanced solid state NMR technique (DIP-HSQC), has been reported [52]. This technique employs REDOR-type recoupling under fast MAS to recouple the heteronuclear ^{1}H-^{15}N dipolar interaction, such that rotor-encoded spinning sideband patterns are obtained.

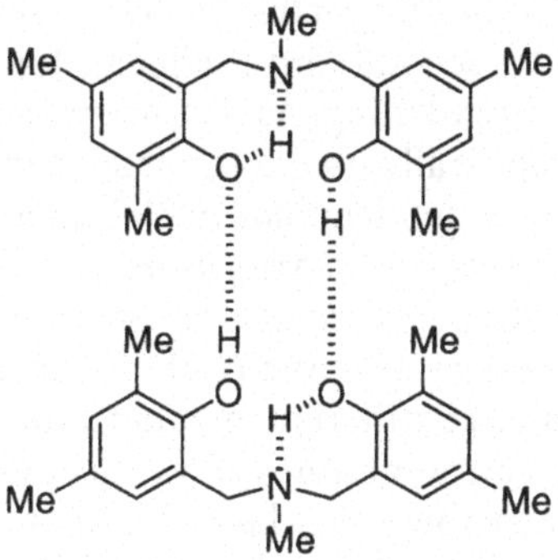

The analysis yields the ^{1}H-^{15}N dipolar coupling and hence the N···H distance. The recoupling scheme used relies on inverse ^{1}H detection which in addition to significant sensitivity enhancement provides better resolution in the ^{1}H dimension. Using this experimental approach, a long N···H distance of 1.94±0.05 Å was determined, which indicates that the proton in the N···H···O hydrogen bond is proximal to the oxygen, while being shared to some extent with the nitrogen. However, the above result was obtained on the assumption of a rigid structure, and the analysis did not include the possible occurrence of proton transfer, and its effects on the dipolar recoupling NMR experiments.

Finally, using both ^{13}C and ^{15}N labelled gramicidin A samples in hydrated phospholipid bilayers, both intermolecular and intramolecular distances have been measured using a solid state NMR technique based on simultaneous frequency and amplitude modulation [53]. By measuring ^{15}N-^{13}C residual dipolar couplings across a hydrogen bond, distances of the order of 4.2±0.2 Å were established.

2.4
^{17}O NMR

Solid state ^{17}O NMR offers another possibility to study hydrogen bonding since in many cases oxygen atoms are directly involved in the hydrogen bond either as donor (OH) or acceptor. The main restriction regarding the use of ^{17}O NMR is the low natural abundance (0.037%) of the ^{17}O isotope. As a result, ^{17}O isotopic enrichment is necessary for solid state NMR studies. In spite of the inconvenience and expense associated with such enrichment, a number of solid state ^{17}O NMR measurements in hydrogen bonded materials, such as L-alanine containing polypeptides [54], benzamide [55], benzoic acid dimer and other organic solids [56], and phthalic acid and its salts [57]. These studies have demonstrated that determination of both the oxygen chemical shift and EFG tensors is possible from the analysis of ^{17}O NMR spectra (both under MAS and for static samples), and both the magnitude and relative orientations of the ^{17}O chemical shift and EFG tensors can be measured. As shown below, these NMR parameters are very sensitive to the hydrogen bonding.

Advantages provided by 2D multiple quantum (MQ) MAS NMR has been used to facilitate the analysis of ^{17}O NMR spectra. As ^{17}O is a spin 5/2 nucleus, ^{17}O MAS NMR spectra are affected by second and higher order quadrupole couplings, which result in severe line broadening. As a consequence, the identification of chemically and crystallographically inequivalent sites in solids may be difficult or impossible. For half-integer quadrupolar nuclei, an asymmetric powder pattern corresponding to the central $\pm 1/2$ transition is usually observed, whereas the satellite transitions are often too broad to be observed. The width of the central powder pattern is inversely proportional to the applied magnetic field strength, and measurements at different magnetic field strengths combined with lineshape simulations can sometimes allow the different species present to be identified. However, when there is a large number of inequivalent species and a relatively small range of isotropic chemical shifts, alternative techniques are required in order to achieve enhanced resolution. The 2D MQ MAS experiment provides an effective separation of the isotropic chemical shifts and anisotropically broadened quadrupolar powder patterns along two dimensions [58, 59]. For example, 2D ^{17}O MQ MAS NMR spectra for four ^{17}O labelled materials [$^{17}O_2$]-D-alanine, potassium hydrogen [$^{17}O_4$]-dibenzoate, hydrochloride of [$^{17}O_4$]-D,L-glutamic acid and [2,4-$^{17}O_2$]-uracil have recently been reported [60]. The high spectral resolution observed in the 2D ^{17}O MQ MAS NMR spectra allowed extraction of precise ^{17}O NMR parameters for all crystallographically distinct oxygen sites.

3
NMR Parameters and Hydrogen Bonding Geometry

3.1
^2H Quadrupole Coupling Constants

Amongst different types of spectroscopic data that may be recorded, vibrational frequencies have been used extensively for correlations with the hydrogen bond distance. The O-H stretching vibration frequencies of non-hydrogen bonded hydroxyl groups are typically in the range 3600–3700 cm^{-1}, whereas for hydrogen bonded hydroxyl groups they are in the range 1500–3600 cm^{-1} [61]. Relationships between O-H stretching frequencies (ν_{OH}) and hydrogen bond distances were first reported in the early 1950s [62, 63]. In earlier work, the correlations were comparatively poorly defined due to the low precision of the crystallographic data. Subsequently, neutron diffraction results were used for correlations with spectroscopic data, including NMR data, leading to significantly improved correlations.

A correlation between the ^2H quadrupole coupling constant e^2qQ/h and O-H stretching frequency ν_{OH} (with e^2qQ/h proportional to ν_{OH}^2) was reported by

Blinc and Hadži [64]. In contrast, however the following linear relationship has been reported [65] for water in solid hydrates:

$$(e^2qQ/h)/\text{kHz} = 0.107\ (\nu_{OH}/\text{cm}^{-1}) - 135 \tag{2}$$

Attempts were also made to correlate the magnitudes of ^{2}H quadrupole coupling constants to hydrogen bond lengths [66–68]. Initially, a $(r_{H\cdots O})^{-3}$ dependence of e^2qQ/h was suggested [67] and an empirical relationship of the form

$$(e^2qQ/h)/\text{kHz} = A - B\ (r_{H\cdots O}/\text{Å})^{-3} \tag{3}$$

was used to fit the quadrupole coupling constants for a variety of hydrogen bond donors and acceptors. In the case of O-D···O interactions, the parameters A=328 and B=643 were derived. However, based on the bond polarisation theory [69] it was suggested instead that the quadrupole coupling constant is proportional to $(r_{H\cdots O})^{-1}$ [70]. This suggestion was confirmed from experimental data for deuterated salts, for which the following relationship was derived:

$$(e^2qQ/h)/\text{kHz} = 560 - 64\ (r_{H\cdots O}/\text{nm})^{-1} \tag{4}$$

In addition, quadrupolar asymmetry parameters (η) were also correlated quantitatively with hydrogen bond geometries [71].

A systematic investigation of methodology for ab initio calculation of ^{2}H quadrupole coupling constants has been reported [72]. The findings of this study, which was focused on the α and β polymorphs of oxalic acid dihydrate, emphasised the importance of considering the full periodic crystal structure in order to obtain ab initio computational predictions in close agreement with experimental values, rather than using just a single molecule or a small cluster of molecules comprising a central molecule and its first shell of hydrogen bonded neighbours. Comparison of the results obtained for these different sizes of system allowed a quantitative assessment of the intramolecular contribution to the ^{2}H quadrupole coupling constant, the intermolecular contribution from the first shell of neighbouring molecules and the intermolecular contribution from outer shells.

Ab initio calculations have also been applied [73] in a systematic study of the geometrical dependence of ^{2}H quadrupole interaction parameters on the geometry of O-H···O=C hydrogen bonds. In this work, the water-formaldehyde complex was used as a model system. Ab initio HF-SCF calculations (using 6-31G** basis set) were carried out as a function of the intermolecular geometry of the complex, leading to an understanding of the dependence of the ^{2}H quadrupole coupling constant and asymmetry parameter on specific geometric parameters defining the hydrogen bonded system.

Correlations between ^{2}H quadrupole interaction parameters and hydrogen bond geometry have also been considered for situations other than O-H···O hydrogen bonds. For example, solid state ^{2}H NMR spectra of ^{2}H labelled amino acids, peptides and polypeptides were measured over a wide temperature range [74]. From spectral simulations based on dynamic ^{2}H NMR theory, parameters such as the ^{2}H quadrupolar coupling constant and asymmetry parameter were determined, and relationships between these NMR parameters and the hydrogen bond distance (in this case $r_{N\cdots O}$) were elucidated. From the observed ^{2}H NMR spectra of amide deuterons of peptides and polypeptides, it was found that the quadrupole coupling constant decreases as $r_{N\cdots O}$ decreases.

3.2 Isotropic ^{1}H Chemical Shifts

Simple correlations have been established between isotropic ^{1}H chemical shifts and O···H and O···O distances in O-H···O hydrogen bonds for a variety of organic and inorganic solids. Correlations between isotropic ^{1}H chemical shift and O···O distance, as well as between ^{2}H quadrupole coupling constant and O···O distance, have also been reported [75].

A nearly linear relationship between the isotropic ^{1}H chemical shift (δ_H) and H···O distance ($r_{H\cdots O}$) has been presented [76] for a series of compounds, using H···O distances determined from neutron diffraction (which are substantially more accurate than those determined from X-ray diffraction). The data were fitted well by a linear plot in which an increase of $r_{H\cdots O}$ by 1.0 Å corresponds to a change of δ_H by −20 ppm. A linear relationship between δ_H and $r_{H\cdots O}$ was found over the whole range studied, from very short (almost symmetrical) hydrogen bonds to long hydrogen bonds (involving water molecules in hydrates). X-ray diffraction data (corrected using a standard value of 0.97 Å for the O-H bond distance) were found to lie on or systematically below the line correlating the neutron diffraction data, suggesting that corrections to the X-ray diffraction data of between 0.98 Å and 1.02 Å would be more appropriate.

A linear relationship between isotropic ^{1}H chemical shift (δ_H) and O···O distance ($r_{O\cdots O}$) has also been established [77] for several metal phosphates and minerals. Similarly, for carboxylic acid protons, δ_H has been shown [78] to depend linearly on $r_{O\cdots O}$, and for several trihydrogen selenites, δ_H was shown [79] to correlate linearly with $r_{O\cdots O}$ and $r_{H\cdots O}$ distances.

Using structural data obtained from neutron diffraction studies for 41 different crystalline solids, the following linear relationship was reported [70]:

$$\delta_H/\text{ppm} = 4.65\ (r_{H\cdots O}/\text{nm})^{-1} - 17.4 \qquad (5)$$

As in the case of ^{2}H quadrupole coupling constants discussed above, this relationship is supported by the bond polarisation theory. Furthermore, a linear relationship between δ_H and the ^{2}H quadrupole coupling constant was reported [70]:

$$\delta_H/\text{ppm} = 26.6 - 0.1\ (e^2qQ/h)/\text{kHz} \qquad (6)$$

In contrast, however, a quadratic relationship between δ_H and e^2qQ/h was used in a recent report [71] based on the earlier correlation of Berglund and Vaughan [75].

In addition, basic quantum mechanical calculations have shown that the change in isotropic ^{1}H chemical shift (δ_H) due to hydrogen bond formation can be attributed primarily to O-H bond polarisation [80]. Similarly, the change in ^{2}H quadrupole coupling constant is also expected to be caused by O-H bond polarisation. It would therefore be interesting to explore correlations between δ_H and the O-H bond length ($r_{O\text{-}H}$) and correlations between $e^2qQ/h$and $r_{O\text{-}H}$, as $r_{O\text{-}H}$

may be expected to be a better indicator of changes in O-H bond polarisation than $r_{H\cdots O}$.

Isotropic ^{1}H chemical shifts for weakly hydrogen bonded hydrates have recently been compared [81] with previous data on carboxylic acids with O-H···O hydrogen bonds of strong and medium strengths. The values of δ_H for the hydrogen bonded protons in this work varied from 4.8 ppm in $NaClO_4 \cdot H_2O$ to 20.5 ppm in potassium hydrogen malonate.

Interestingly, an extreme value of isotropic ^{1}H chemical shift (δ_H=–4.4 ppm) has been reported [82] for the O-H···O hydrogen bonding in solid KOH. The zigzag chains of oxygen atoms in the crystal structure of KOH [83] are linked by weak hydrogen bonds with $r_{H\cdots O}$=2.776 Å and ∠O-H···O=155°. Interestingly, the hydrogen bonded protons in KOH appear to be even more shielded than those in water vapour (δ_H=1.2 ppm), in which hydrogen bonding is essentially absent.

A clear correlation between isotropic ^{1}H chemical shift and the frequency of the O-H stretching vibration has been reported [61] for surface hydroxyl groups in zeolites and related materials, as well as for water molecules in solid hydrates and strongly hydrogen bonded protons in inorganic solids.

Correlations between isotropic ^{1}H chemical shift and hydrogen bonding geometry have also been reported for situations other than O-H···O hydrogen bonding. For example [84], the values of δ_H for the Gly amide protons of Gly-containing peptides and polypeptides have been shown to move more downfield as the N···O distance in the N-H···O hydrogen bonding decreases.

3.3 ^{1}H Chemical Shift Anisotropy

Clearly the chemical shift anisotropy may be a more detailed source of structural and dynamic information than the isotropic chemical shift. Early work demonstrated that ^{1}H CSA measurements are more sensitive to structural changes than ^{1}H isotropic chemical shift measurements. However, understanding the ^{1}H CSA and its dependence on hydrogen bond geometry has been rather controversial. In the initial publications, correlation of the hydrogen bond geometry and ^{1}H chemical shift anisotropy, Δ, was considered to be less straightforward as the latter is strongly influenced by through-space shielding effects. Another factor that might complicate the interpretation of CSA measurements is the effect of motional averaging on Δ, as both small-angle and large-angle reorientations are likely to decrease the measured Δ value, whereas the motion may have little or no effect on the isotropic chemical shift. Hence, measurements of Δ at very low temperatures should be more suitable for correlations with hydrogen bond geometry. Nevertheless, it was found that the formation of hydrogen bonds generally leads to an increase in ^{1}H CSA [76, 85] although, unlike δ_{iso}, the relationship between the CSA and H···O distance was more scattered. Clearly, a better approach would be to correlate individual components of the ^{1}H chemical shift tensor with geometric parameters describing the hydrogen bond geometry (clearly this requires the directions of the ^{1}H CSA components to be determined). Numerous experimental studies have shown that for hydrogen bonded protons in OH groups, the ^{1}H chemical shift tensor is axially symmetric with the principal axis lying along the vector joining the hydrogen bonded atoms [7]. A relatively large number of ^1H CSA measurements have been reported [86]. As an illustration of the magnitude of the hydrogen bonded CSAs, the principal components of the chemical shift tensor for β-oxalic acid are 21.6, 21.6 and –4.2 ppm [82]. On the assumption that the ^1H chemical shift tensor has components $\delta_{||}$ and $\delta_{\perp}$ parallel and perpendicular to the O-H bond direction respectively [75], correlations were made with the ^2H quadrupole coupling constant e^2qQ/h (which provides a reliable measure of the strength of the hydrogen bond). It was found that $\delta_{\perp}$ varies significantly with e^2qQ/h, whereas no significant variation was found for $\delta_{||}$. The chemical shift tensor component $\delta_{\perp}$ and the isotropic chemical shift δ_{iso} were also found to correlate linearly with O···O distance for O-H···O hydrogen bonds of moderate and high strengths [87]. By extending the dataset to include weakly hydrogen bonded solids, such as KOH, a new empirical relationship correlating $\delta_{\perp}$ and δ_{iso} with $exp(-r_{O\cdots O}/\rho)$ has been suggested (with ρ= 0.94 Å) [82]. The specific functional form used for this correlation originates from an interpretation of O-H···O hydrogen bonding in terms of a simple ionic model.

Further revision of the above correlations for ^1H CSA have been carried out under the recognition that the assumption about the axiality of the ^1H chemical shift tensor is not always true. In addition, the fact that only one ^1H chemical shift tensor component showed significant dependence on the hydrogen bond distance [75] did not agree with the earlier theoretical predictions. In particular, ab initio calculations on the water dimer have demonstrated that hydrogen bonding affects the ^1H chemical shift tensor by two principal mechanisms [80]: (i) the "electron depletion effect", which is an essentially isotropic deshielding resulting from the reduced electron density on the hydrogen atom, and (ii) the "acceptor effect", which describes the effect at the proton site generated by the electron distribution at the acceptor oxygen. The latter effect shifts $\delta_{||}$ and $\delta_{\perp}$ in opposite directions, and may therefore be expected to dominate the dependence of $\Delta=\delta_{||}-\delta_{\perp}$ on hydrogen bonding. On the other hand, δ_{iso} depends on both the acceptor effect and the electron depletion effect [80]. These and other theoretical results [88] indicate that variations in hydrogen bond geometry should be more strongly manifested in the CSA than in the isotropic chemical shift, in accordance with the recent experimental findings.

A significantly improved experimental study was undertaken recently [81]. In accordance with earlier recommendations [78], a set of closely related solids was chosen in order to reduce data scatter. In particular, weakly hydrogen bonded water molecules in magnetically ^1H dilute crystalline hydrates were used for ^1H chemical shift tensor measurements and for hydrogen bond correlations. It was found that the most shielded and least shielded components of the ^1H chemical shift tensor change in opposite directions as a function of the hydrogen bond distance. Hence, it was confirmed that ^1H CSA is a more sensitive measure of hydrogen bond strength than ^1H isotropic chemical shift.

For example, over the range of H···O distances from 1.66 Å to 2.15 Å, the span of the ^1H chemical shift tensor [$\Omega=\delta_n-\delta_{||}$, where δ_n is the ^1H chemical shift tensor component normal to the H_2O plane and $\delta_{||}$ is the in-plane component parallel to the O···O vector (Fig. 3)] changes by more than 20 ppm, and is nearly

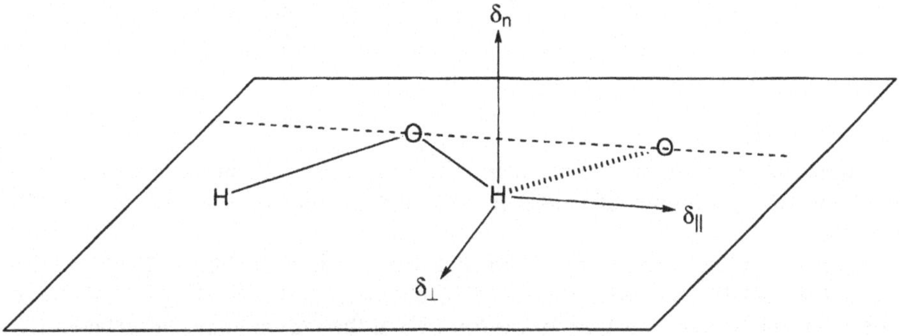

Fig. 3 ^{1}H chemical shift tensor orientations for a static water molecule [81]. The ^{1}H chemical shift tensor component δ_n is normal to the H_2O plane. In-plane components $\delta_{\|}$ and $\delta_{\perp}$ are parallel and perpendicular (respectively) to the O···O hydrogen bond direction

four times higher than the change in the isotropic ^{1}H chemical shift. An approximately linear relationship was reported for the dependence between Ω and $r_{H\cdots O}$ with a correlation coefficient of ca. 0.95.

3.4 Isotropic ^{13}C Chemical Shifts and ^{13}C Chemical Shift Anisotropy

It is well known [89] that, in solution state ^{13}C NMR, the ^{13}C chemical shift for C=O carbons is shifted to a higher value by hydrogen bonding. In general, this is also observed for the isotropic ^{13}C chemical shift in solid state ^{13}C NMR. A classic illustration is α-diacetamide, the crystal structure [90] of which contains dimers, with only one of the two carbonyl groups involved in hydrogen bonding:

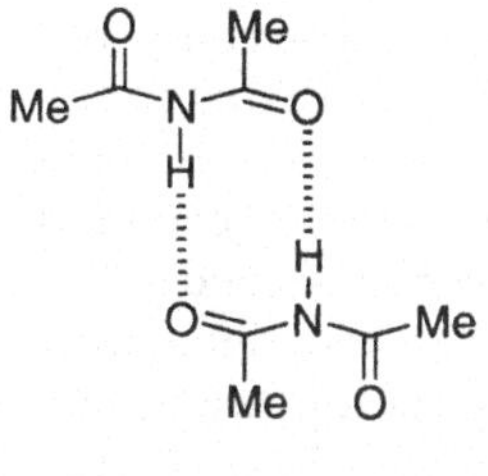

This is observed directly in the high-resolution solid state ^{13}C NMR spectrum by the fact that there are two peaks, separated by 6 ppm, due to carbonyl carbons. The peak at higher frequency is assigned as ^{13}C in the hydrogen bonded carbonyl group [91, 92].

Another simple example concerns the ability of high-resolution solid state ^{13}C NMR to distinguish different conformations of symmetrically substituted acyclic imides [92]. In the *cis-trans* conformation, the two carbonyl groups can be distinguished (as for α-diacetamide), whereas for the *trans-trans* conformation, the two carbonyl groups have similar hydrogen bonding and crystallographic environments and are indistinguishable by ^{13}C NMR.

Thus, the *cis-trans* and *trans-trans* conformations can be distinguished on the basis of the number of peaks in their high-resolution solid state ^{13}C NMR spectra.

High-resolution solid state ^{13}C NMR studies of hydroxybenzaldehydes have probed the relation between isotropic ^{13}C chemical shift and O···O distance for O-H···O hydrogen bonds [93]. It was found that the ^{13}C chemical shift (taken relative to the chemical shift for the same molecule in DMSO-d_6 solution) of the aldehyde carbon increases as a result of intramolecular or intermolecular hydrogen bonding, and the increase varies inversely with the O···O hydrogen bond distance determined from X-ray diffraction (i.e. the smaller the O···O distance, the larger the chemical shift difference between the molecule in the solid state and the same molecule in DMSO-d_6 solution). When there is no possibility of hydrogen bonding, the chemical shift difference is only ca. 0.4 ppm (in comparison with ca. 2–7 ppm when hydrogen bonding exists). No attempt was made to correlate the O···O distance and the strength of the hydrogen bonds, as the shortest O···O distance occurs for an intramolecular hydrogen bond which need not be any stronger than a longer, more linear, intermolecular hydrogen bond.

A detailed study of intermolecular hydrogen bonding effects has been based on determination of the ^{13}C chemical shift tensor for the carbonyl carbon in a single crystal of dimedone (5,5-dimethyl-1,3-cyclohexanedione) [94]. The ^{13}C NMR chemical shifts for carbonyl and enol carbons in solid dimedone are higher than for the same molecule in DMSO-d_6 as a consequence of intermolecular hydrogen bonding in the solid. The complete ^{13}C chemical shift tensor was determined for the carbonyl group from single crystal ^{13}C NMR spectra recorded as a function of crystal orientation. It is interesting to compare the ^{13}C chemical shift tensor for the carbonyl carbon in dimedone with that in acetophenone, which does not engage in intermolecular hydrogen bonding. In both cases, the carbonyl group is bonded to sp^2 and sp^3 carbons, so the ^{13}C chemical shift tensors can be compared directly and any differences can be assigned to the presence of intermolecular hydrogen bonding in dimedone. It is found that the hydrogen bonding causes small variations in electronic configuration resulting in the apparent downfield shift (ca. 50 ppm) of the δ_{22} component of the ^{13}C chemical shift tensor of dimedone compared with acetophenone. This downfield shift of the δ_{22} component is the main contributor to the well established downfield shift of isotropic ^{13}C chemical shifts observed for hydrogen bonded carbonyl groups in high-resolution ^{13}C NMR spectra of solids [95, 96].

The most detailed correlation between the strength of hydrogen bonding and the carbonyl ^{13}C CSA has been reported for peptides, for which the most shielded component δ_{33} is perpendicular to the C_α-C(O)-N plane, the component δ_{22} lies approximately along the C=O bond, and the least shielded component δ_{11} is approximately parallel to the direction of the C_α-C(O) bond (Fig. 4) [97–99].

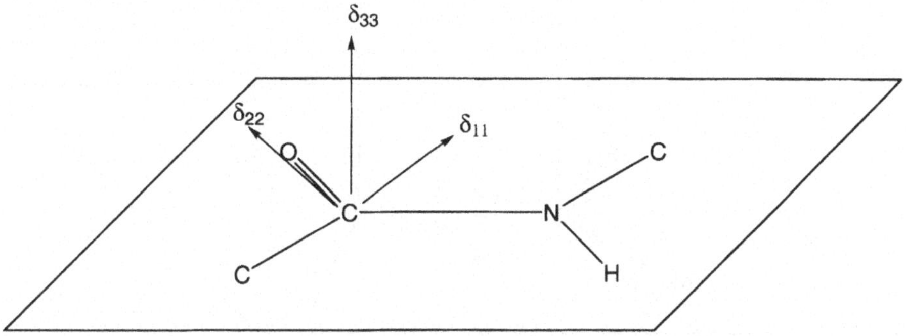

Fig. 4 Schematic presentation of the ^{13}C CSA tensor orientation in peptides

On the basis of solid state NMR studies on single crystals [97, 100] and powder samples [98, 101–103] of peptides, it has been shown that a large high-frequency shift for δ_{22}, a low-frequency shift for δ_{11} and no change for δ_{33} are expected to result from a decrease in the N···O distance ($r_{N\cdots O}$) in N-H···O hydrogen bonds. It has been shown that the range of isotropic chemical shifts δ_{iso} for the carbonyl carbons in proteins predominantly arises from the dependence of δ_{22} on the secondary structure [104]. Linear relationships between δ_{iso} and $r_{N\cdots O}$ have been established for the carbonyl carbons of a number of amino acid residues in peptides and polypeptides in the crystalline state [101, 104, 105]. Based on these linear relationships and assuming that the conformation dependent ^{13}C chemical shift of the amide carbonyl is caused by the change in the hydrogen bond distance, ^{13}C chemical shift contour maps were constructed as a function of the dihedral angles (φ,ψ) in the vicinity of the α-helix conformation [106]. The dependence of the C_α and C_β CSA components in polypeptides on the conformation and dynamics of the side chain as well as on the packing interactions and the secondary structure have also been reported [107]. This work has also demonstrated the potential of the 2D PASS (phase adjusted spinning sideband) technique developed [108, 109] for the measurement of CSA components.

3.5 Isotropic ^{15}N Chemical Shifts and ^{15}N Chemical Shift Anisotropy

Similar to the situation for ^{13}C, isotropic ^{15}N chemical shifts and the principal components of ^{15}N chemical shift tensors have been used to study N-H···O=C hydrogen bonds in peptides. It has been shown that isotropic ^{15}N chemical shifts of proton donors (such as N-H) are displaced downfield by ca. 15 ppm, whereas those of proton acceptors are shifted upfield by ca. 20 ppm [110–112]. Amongst the CSA components, δ_{33} (parallel to the C-N bond) has been shown to be most sensitive to the hydrogen bond strength, as reflected by the N···O distance [113]. Detailed studies of the principal components and orientations of ^{15}N chemical shift tensors for amide nitrogens in simple peptides have been reported recently [114]. This work confirmed that δ_{33} and δ_{iso} are the ^{15}N chemical shift parameters that are the most sensitive to details of the hydrogen bonding. It was also found that N-H

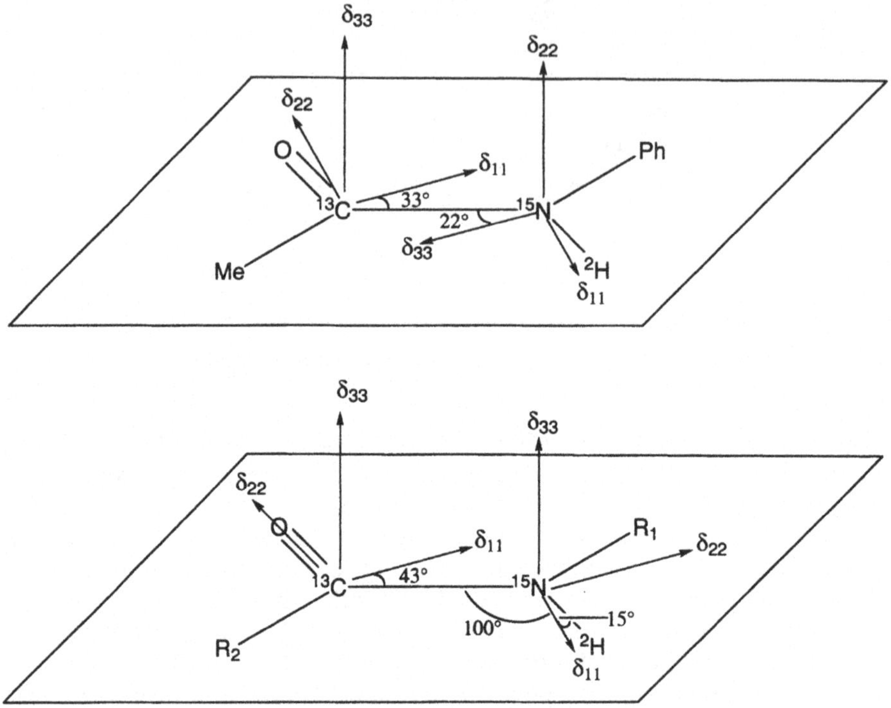

Fig. 5 Schematic representation of the ^{13}C and ^{15}N chemical shift tensor orientations in acetanilide (*top*) and gluconamide (*bottom*) [117]

bond lengths calculated from the measured ^{15}N-^{1}H dipole-dipole interaction could be overestimated in some cases when the amplitude of thermal motion of the N-H moiety is significant, as encountered in weakly hydrogen bonded solids.

Several recent studies have emphasised the advantages of combining the analysis of CSA and dipolar interactions in studies of hydrogen bonded systems [115, 116]. Although dipole-dipole interactions are a good source of information on internuclear distances, these interactions have axial symmetry and information on the relative orientations of the interacting nuclei is not accessible from measurements of dipole-dipole interactions. Orientational information can instead be assessed from CSA measurements. For example, for the amide bond fragment, the most shielded component of the ^{15}N chemical shift tensor is along the direction of the ^{13}C-^{15}N bond, and the component of intermediate magnitude is perpendicular to the plane of the amide group. Using this combined approach, the hydrogen bond structures in gluconamide fibres have been studied [117]. The ^{13}C-^{15}N dipolar interaction was determined from the SEDOR experiment [118], and the combined dipolar-chemical shift NMR approach was used to correlate the ^{15}N and ^{13}C chemical shift tensors, allowing the relative orientations of the two chemical shift tensors to be determined. Some major differences were found with regard to the orientations of the ^{13}C and ^{15}N chemical shift tensors in the amide plane for gluconamide and acetanilide, schematic representations of which are shown in Fig. 5.

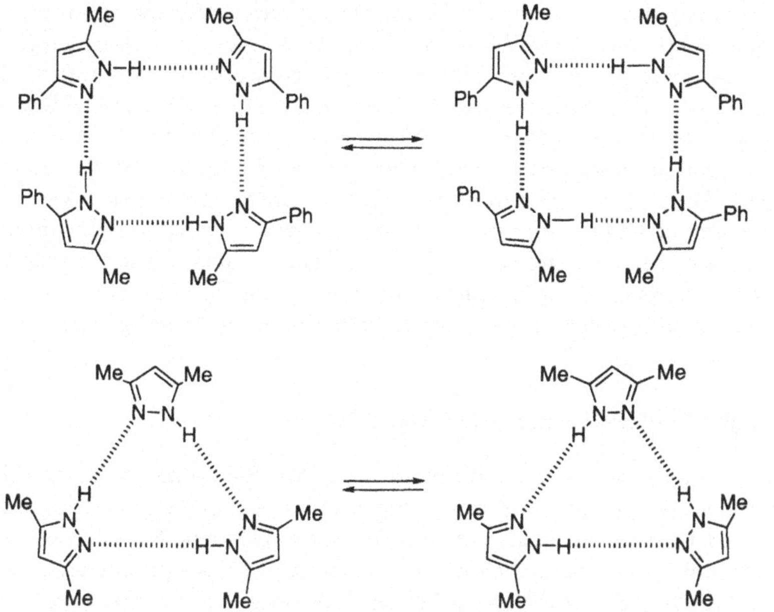

Fig. 6 Degenerate proton transfer in cyclic tetramers of crystalline 5-methyl-3-phenylpyrazole (*top*) and in cyclic trimers of crystalline 3,5-dimethylpyrazole (*bottom*) [122]

The observed differences have been attributed to differences in the intermolecular N-H···O=C hydrogen bonding, which is the main contributor to the formation of extended quadruple helices in the case of the fibrous gluconamide.

^{15}N dipolar interactions and chemical shifts have also been used to study hydrogen bonded structures found in enzymes. Changes in hydrogen bonding interactions between ground states and transition states can make an important contribution to enzyme catalysis [119], and understanding the hydrogen bonded structures is crucial for the development of new artificial enzymes. Advanced solid state NMR techniques have been employed to model hydrogen bonded complexes in enzymes. Using the 2D 2φ-DipShift technique [120], N-H bond lengths in imidazolium-carboxylate pairs have been determined [121]. Histidine complexes were chosen in order to model enzyme active sites. The technique used relies on determining ^{15}N-^{1}H heteronuclear dipolar coupling *via* numerical simulations of heteronuclear dipolar spinning sideband patterns. The values measured for various compounds were in the range from 1.013 Å (corresponding to $r_{N\cdots O}$=2.716 Å and ∠NHO=134°) to 1.103 Å (corresponding to $r_{N\cdots O}$=2.933 Å and ∠NHO=170°). It was found that both ^{1}H and ^{15}N chemical shifts occur further downfield in the case of longer N-H bonds. In addition, an almost linear correlation was noted between isotropic ^{15}N chemical shift and N-H bond length.

The geometries of hydrogen bonded trimers and tetramers in solid 3,5-substituted pyrazoles (Fig. 6) have been studied from consideration of both ^{15}N chemical shift tensors and dipolar interactions involving ^{15}N [122].

The principal values of the ^{15}N chemical shift tensors for the amine and imine nitrogen atoms were derived from analysis of ^{15}N lineshapes recorded for static powder samples under conditions of ^{1}H-^{15}N cross polarisation and ^{1}H decoupling, and the orientations of the ^{15}N CSA components in the molecular principal axis system were obtained by taking into account the ^{15}N-^{15}N and ^{15}N-^{2}H dipole-dipole interactions (the latter for selectively deuterated materials). The relative orientations of the amine and imine chemical shift tensors were also independently checked using off-MAS magnetization transfer experiments. For both types of nitrogen atom, the isotropic ^{15}N chemical shifts, the magnitudes and orientations of the principal components of the chemical shift tensors, and the N-D distances depend only slightly on hydrogen bonding geometry.

3.6
^{17}O Chemical Shift and Electric Field Gradient Tensors

The dependence of ^{17}O NMR parameters on hydrogen bonding geometries has been less studied, mainly due to the unfavourable NMR characteristics of the ^{17}O nucleus that render such experiments difficult to perform. Nevertheless, a number of recent reports have shown that solid state ^{17}O NMR parameters are sensitive to the strengths of hydrogen bonds. Some examples of these recent developments are given below.

The ^{17}O EFG and chemical shift tensor components, and their relative orientations, have been reported for the carbonyl groups in crystalline benzamide [55], other solid amides [123], urea [124] and nucleic acid bases [125]. Urea presents a unique example of hydrogen bonding, as each carbonyl oxygen atom is involved in hydrogen bonding with four different N-H bonds. As reported previously [126], there is no large-angle motional averaging in solid urea at 303 K, thus making urea an ideal candidate for studying hydrogen bonding effects on the NMR parameters. The ^{17}O quadrupole coupling constant and asymmetry parameter in crystalline urea were found to be 7.24 MHz and 0.92 respectively, and the principal components of the ^{17}O chemical shift tensor were determined to be δ_{11}=300 ppm, δ_{22}=280 ppm and δ_{33}=20 ppm. The principal component with the lowest shift δ_{11} is perpendicular to the C=O bond and the principal component with the highest shift δ_{33} is perpendicular to the molecular plane. Quantum mechanics calculations revealed that intermolecular hydrogen bonding has a large effect on the ^{17}O NMR tensors, and that the ^{17}O quadrupole coupling constant decreases as the number of hydrogen bonds is increased. These calculations also showed that the presence of the four C=O···H-N hydrogen bonds in crystalline urea causes a decrease of 1 MHz in the ^{17}O quadrupole coupling constant and an increase of 50 ppm in the isotropic ^{17}O chemical shift. It was also demonstrated that inclusion of a complete intermolecular hydrogen bonding network is necessary in order to obtain reliable ^{17}O EFG and chemical shift tensors and calculations with a molecular cluster comprising seven urea molecules yielded ^{17}O NMR tensors in reasonably good agreement with the experimental data.

Recent ^{17}O NMR experiments on phthalate species has confirmed that both the ^{17}O isotropic chemical shift and ^{17}O quadrupole coupling constant decrease as hydrogen bond strength increases [57]. These experimental observations have been supported by results from theoretical studies [127].

Some aspects of the dependence of ^{17}O NMR parameters on the geometry of hydrogen bonding have been explored in experimental studies of polyglycines [128], poly(L-alanine)s [54] and other peptides and polypeptides [129]. It was found that the isotropic ^{17}O chemical shifts for carbonyl groups in polypeptides move upfield as the N···O distance in N-H···O hydrogen bonds decreases, and the ^{17}O quadrupole coupling constant decreases as the N···O distance decreases. Differences in the chemical shift values between peptides and polypeptides were attributed to differences in molecular packing.

Theoretical calculations employing density functional theory have been carried out to determine ^{17}O quadrupole coupling constants and asymmetry parameters for small α-helix and β-sheet protein fragments [130]. It was found that the ^{17}O quadrupole parameters of proteins depend on the conformation of the backbone, and specifically on the hydrogen bond angle ∠H-N···O and the backbone dihedral angle ∠NC-C(O)N. For this reason, ^{17}O quadrupole interaction parameters show observable differences between α-helices and β-sheets. Interestingly, it was found that ^{17}O quadrupole coupling constants do not depend on the hydrogen bond distance, and do not depend on either the hydrogen bond dihedral angle ∠N-C=O···H or the backbone dihedral angle ∠C(O)C-NC(O).

3.7 Overview

Overall, various studies have shown that NMR parameters relating to chemical shift and quadrupole interactions for various types of nucleus in the vicinity of hydrogen bonds often correlate well with parameters describing the hydrogen bond geometry. In particular, studies of solids containing O-H···O hydrogen bonds have reliably shown that an increase in ^{1}H isotropic chemical shift and a decrease in ^{2}H quadrupole coupling constant correspond to decreases in both the O···O and O···H distances. Following theoretical predictions, it has also been shown that ^{1}H CSA is more sensitive to changes in hydrogen bond geometry than the isotropic ^{1}H chemical shift. As the O···O and O···H distances determined by diffraction techniques are generally interpreted as a direct measure of the strength of the hydrogen bond, these NMR parameters can be used to provide an indication of the strengths of hydrogen bonds. This approach is especially suitable when comparison of NMR parameters is made for well-defined families of materials that are related both in terms of structure and dynamics. In comparison with measurements of stretching vibration frequencies, the advantage of using NMR parameters is that the experimental errors of NMR measurements are normally less than that of IR measurements since IR vibration bands are often broadened considerably by hydrogen bonding. Although high-resolution solid state ^{1}H NMR spectra can be difficult to obtain, identification of the isotropic ^{1}H chemical shifts for hydrogen atoms involved in hydrogen bonding may be possible from spectra of moderate resolution, as the isotropic ^{1}H chemical shifts of such hydrogens are often higher than those for other types of hydrogen atoms. The availability of new fast MAS probes allowing sample spinning at MAS frequencies up to 35 kHz, as well as new 2D techniques that allow resolution of the ^{1}H NMR spectrum along the second dimension, further assists the measurement of isotropic ^{1}H chemical shifts for initial characterization of hydrogen bonded systems.

In terms of practicality, the correlations of the type discussed above are very useful, but can generally be used only at a semi-quantitative or qualitative level. It is over-optimistic to expect that accurate hydrogen bond distances can be determined for materials of unknown structure using known correlations between NMR parameters and hydrogen bond geometry, as the NMR parameters depend on a number of structural and dynamic factors that may differ from one family of materials to another. Nevertheless, such correlations, combined with theoretical studies, are of considerable importance for unravelling the main structural factors that govern hydrogen bonding interactions.

4 Examples of Applications

4.1 Structural Aspects of Hydrogen Bonding Arrangements in Solids

4.1.1 *Carboxylic Acids*

Detailed studies of carboxylic acids have been carried out using ^{1}H, ^{2}H, ^{13}C and ^{17}O NMR, which have greatly contributed towards the understanding of both structure and dynamics of hydrogen bonding in these solids. Among early work, maleic acid was used to illustrate the relationship between the isotropic ^{1}H chemical shift (δ_{iso}) and hydrogen bond length ($r_{H\cdots O}$) [7]. It was found that for short O···O distance, δ_{iso} is shifted downfield. There are two different types of hydrogen bond in solid maleic acid, with $r_{O\cdots O}$=2.502 Å (an intramolecular hydrogen bond) and $r_{O\cdots O}$=2.643 Å (an intermolecular hydrogen bond). Maleic acid is the *cis* isomer of ethylene dicarboxylic acid, whereas the *trans* isomer is fumaric acid.

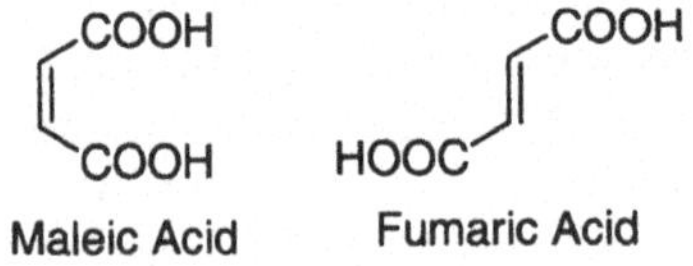

The ^{1}H NMR spectrum for fumaric acid contains two lines, assigned to the olefinic and carboxylic acid protons, the latter of which is characterised by the same O···O distance and has the same δ_{iso} value as the intermolecular hydrogen bond in maleic acid.

A single crystal ^{1}H NMR study of potassium hydrogen maleate has established the chemical shift tensors of all magnetically inequivalent ^{1}H nuclei in the unit cell [131].

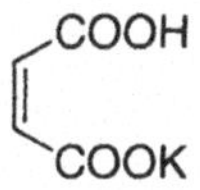

Potassium Hydrogen Maleate

The orientation of the ^{1}H chemical shift tensor for the carboxylic acid group was found to be consistent with the position of the hydrogen atoms at the mid-points between the two oxygen atoms in the hydrogen bond.

Recently a novel series of hydrogen bonded 1:1 acid-base complexes between ^{15}N-labelled 2,4,6-trimethylpyridine (collidine) and carboxylic acids (including derivatives deuterated in the carboxylic acid group) have been studied by ^{1}H MAS NMR and ^{15}N CP NMR with and without MAS [132]. Zwitterionic complexes with the hydrogen bonded proton closer to nitrogen than to oxygen, as well as molecular complexes with the proton located closer to oxygen, were observed [133]. Two of the five complexes studied, with a different location of the hydrogen bonded proton, are shown below.

For these complexes, the isotropic ^{1}H and ^{15}N chemical shifts and the ^{15}N chemical shift tensor elements were measured as a function of the hydrogen bond geometry. Lineshape simulations of the static powder ^{15}N NMR spectra revealed the dipolar ^{2}H-^{15}N couplings and hence the corresponding distances. The results revealed several correlations between hydrogen bond geometry and NMR parameters which were analysed in terms of the valence bond order model. It was shown that the isotropic ^{15}N chemical shifts of collidine and other pyridines depend in a characteristic way on the N-H distance. A correlation of the ^{1}H and ^{15}N isotropic chemical shifts was observed which agrees well with the previously established correlation in which the A···B distance in an A-H···B hydrogen bond decreases significantly when the proton is shifted to the centre of the hydrogen bond.

4.1.2
Peptides and Amides

Solid state ^{2}H NMR has been used to obtain detailed structural information for the amide and carboxylic acid hydrogen sites in a single crystal of the model peptide *N*-acetyl-D,L-valine [134]. Both the amide and carboxylic acid hydrogens are involved in intermolecular hydrogen bonds. The results were compared with experimental data obtained for acetylanilide [135] and ab initio calculations for glycylglycine [136].

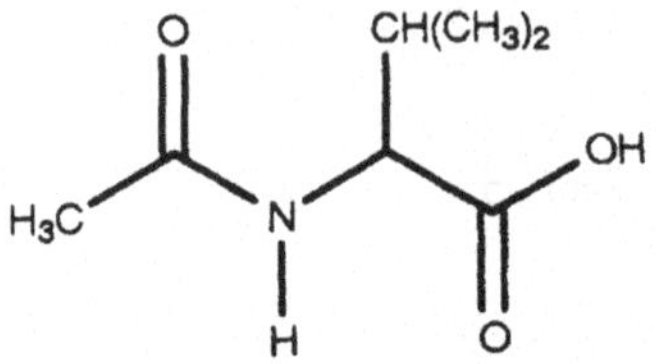

N-Acetylvaline

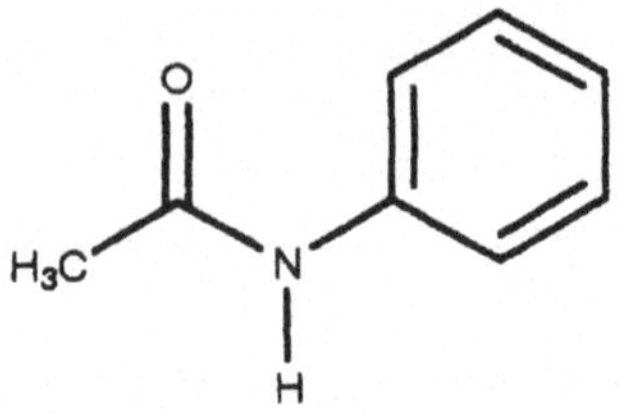

Acetanilide

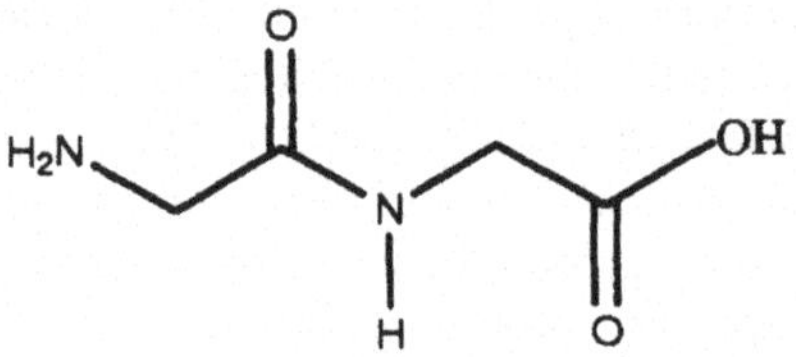

Glycylglycine

Both the magnitudes and directions of the ^{2}H EFG and chemical shift tensors were fully characterised. The quadrupole coupling constant was found to be only 160.3 kHz for the carboxylic acid deuteron, corresponding to a moderately strong intermolecular hydrogen bond with $r_{O\cdots O}$=2.62 Å. The larger quadrupole coupling constant of 212.6 kHz for the amide deuteron was consistent with the weak nature of the intermolecular hydrogen bond in this case, with $r_{N\cdots O}$=3.18 Å. The chemical shift tensor for the amide deuteron in *N*-acetyl-D,L-valine was consistent with the results obtained experimentally for acetanilide and the ab initio calculations for glycylglycine. These results suggest that there is a close correlation between the strength of the N-H···O hydrogen bond and the values of δ_{iso} and Δ, similar to the well established correlation for O-H···O hydrogen bonds.

^{2}H EFG and chemical shift tensors for all the exchangeable deuteron sites in the model dipeptide glycylglycine monohydrochloride have been determined [71].

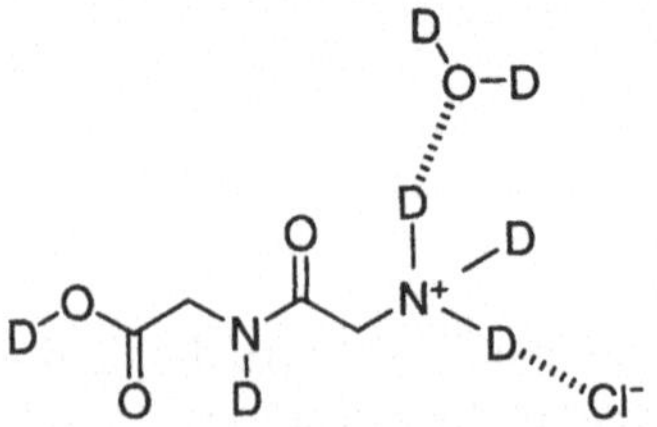

For all three sites at room temperature, the principal axis corresponding to the largest component of each EFG tensor lies nearly along the appropriate bond. Specifically, the carboxylic acid deuteron tensor deviates by only ca. 3° from the bond orientation established by neutron diffraction, while the other two sites are within 1° of the bond orientations. The magnitudes of the quadrupole coupling constants were found to agree well with the empirical relationship given by Eq. (3). The orientations of the ^{2}H chemical shift tensors were also found to correlate with the molecular geometry. The principal axes of the chemical shift tensor for the amide deuteron were all within 2° of the EFG tensor axes. Interestingly, the unique axis (most shielded) of the chemical shift tensor for the OD site was found to form angles of ca. 11° with the O-D bond and 6.5° with the O···O vector. These deviations are attributed to the non-linearity of the hydrogen bonds.

The nature of the hydrogen bonding in polymorphs of *N*-benzoyl-D,L-phenylalanine and *N*-benzoyl-L-phenylalanine has been investigated by solid state ^{13}C NMR [137].

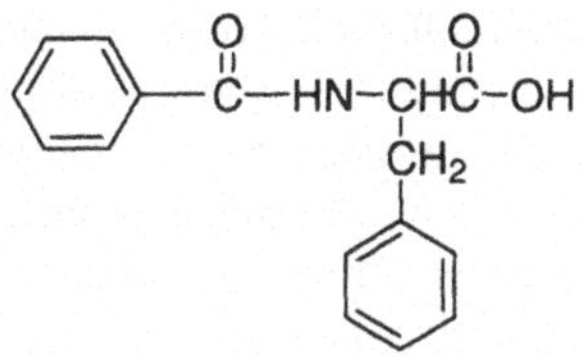

N-Benzoyl-D,L-Phenylalanine

The multiple resonances observed for the carbon of the carboxylic acid group in *N*-benzoyl-L-phenylalanine were shown to be related to different types of hydrogen bonding. These results are in good agreement with earlier studies using ^{1}H CRAMPS NMR [138]. The differences in the intermolecular distances of the carboxylic acid groups involved in different types of hydrogen bonding have been visualised using ODESSA (one-dimensional exchange spectroscopy by sideband alteration) and 2D EXSY (exchange spectroscopy). The ODESSA technique [139] can measure internuclear distances (up to 9 Å) between chemically equivalent nuclei with the same isotropic chemical shift. Potential applications of this approach are widespread.

The ionisation state and hydrogen bonding environment of the transition state analogue inhibitor, carboxymethyldethia coenzyme A, bound to citrate synthase have been investigated using solid state NMR [140]. The enzyme-inhibitor complex was studied in connection with the postulated contribution of short hydrogen bonds to binding energies and enzyme catalysis. The crystal structure of this complex [141] has an unusually short hydrogen bond between the carboxylate group of the inhibitor and an aspartic acid side chain. To further investigate the nature of this short hydrogen bond, ^{13}C chemical shift tensor values describing the CSA of the carboxylic acid group of the inhibitor were obtained (233, 206 and 105 ppm). Comparison of these values with previously reported data and ab initio calculations of ^{13}C chemical shift tensors clearly indicates that the carboxylic acid group is deprotonated. Overall, solid state ^{1}H and ^{13}C NMR studies were in agreement with the suggestion that a very short hydrogen bond is formed.

Accurate ^{13}C-^{15}N interatomic distances have been measured by means of rotational echo double resonance (REDOR) experiments for oligopeptides [50]. The interatomic ^{13}C-^{15}N distance in the hydrogen bonded fragment was measured to be 4.5±0.1 Å in five different samples studied. This finding is consistent with an α-helix structure, in agreement with conformation-dependent ^{13}C chemical shift data.

High-resolution solid state ^{31}P NMR has been applied to probe hydrogen bonding patterns in co-crystals of amides and triarylphosphine oxides [142]. Triarylphosphine oxides form hydrogen bonded co-crystals with a wide range of molecules containing hydrogen bond donor groups. In these materials, the Ar_3PO molecules can form one, two or three N-H···O=P hydrogen bonds per Ar_3PO molecule. It was shown that there is a linear correlation between the isotropic ^{31}P chemical shift (δ_P) and the number of hydrogen bonds (n_H) per Ar_3PO molecule in these co-crystals (Fig. 7).

The dominant factor in the correlation between δ_P and n_H appears to be the expected deshielding of the ^{31}P nucleus as the number of hydrogen bonds increases, and other environmental factors appear to have comparatively little effect. Interestingly, the P-O distances are essentially the same in all the systems considered, and are independent of n_H. This correlation was utilised to predict that $n_H=1$ for the 1:1 co-crystal of unknown structure between Ph_3PO and HN(COMe)Ph.

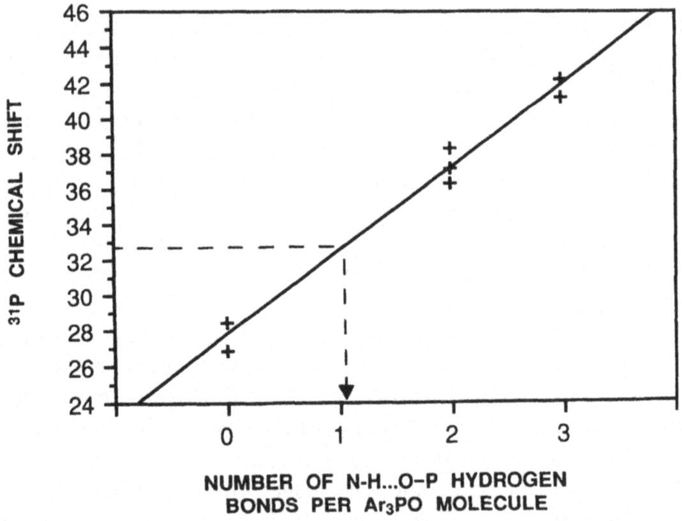

Fig. 7 Graph showing the linear correlation between ^{31}P chemical shift (in ppm) and the number of hydrogen bonds N-H···O=P per Ar_3PO molecule [142]. Experimental data points for cocrystals of known structure are denoted +. The best fit straight line through these points is shown

4.1.3
Other Examples

The high-resolution solid state ^{13}C NMR spectra of polymorphs of naphthazarin (5,8-dihydroxy-1,4-napthoquinone) and methyl derivatives have been rationalised in terms of hydrogen bonding effects [143]. There are three polymorphs (denoted A, B and C) of naphthazarin, in each of which, at room temperature, the molecules occupy crystallographic inversion centres. As shown below, there is a proton transfer reaction between the two forms 1a and 1b. In solution, this proton transfer is fast on the ^{1}H and ^{13}C NMR timescales.

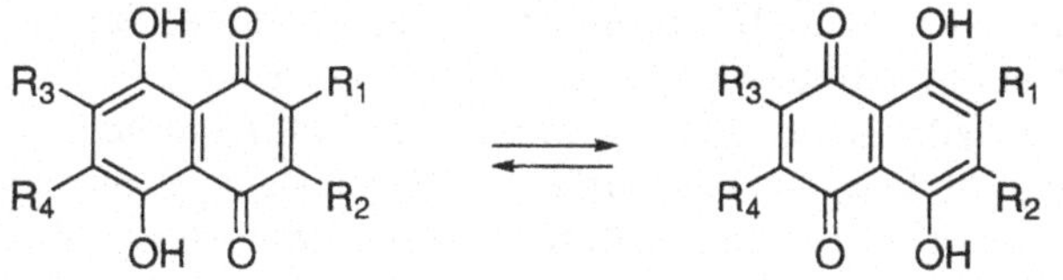

1a 1b

High-resolution solid state ^{13}C NMR spectra were recorded at room temperature for all three polymorphs. The spectra for polymorphs A and B were very similar, but the spectrum for polymorph C was significantly different, as a consequence of differences in the hydrogen bonding arrangements. For polymorphs A and B, the main interaction is C-H···O hydrogen bonding, whereas polymorph C has O-H···O hydrogen bonding, which significantly affects the ^{13}C NMR resonances of the carbons bonded to these oxygen atoms. The high-resolution solid state ^{13}C NMR spectrum of 2,7-dimethylnaphthazarin is similar to that recorded in solution, consistent with fast proton exchange between tautomeric structures with nearly equal populations at room temperature. These results were further supported by ^{1}H spin-lattice relaxation time measurements of naphthazarin A and ^{2}H spin-lattice relaxation time measurements of deuterated hydroxyl groups in naphthazarin C [144]. The results were interpreted in terms of a relaxation model in which the proton or deuteron jumps between two potential minima in the vicinity of adjacent quinonoid and hydroxyl oxygens. The low temperature relaxation data were interpreted in terms of a model in which quantum mechanical tunnelling dominates, whereas the relaxation rates at higher temperatures were explained by classical jumps across the barrier of the asymmetric potential well.

Next, we consider the application of ^{13}C NMR to probe materials containing N-H···N hydrogen bonds. In the crystal structure of campho[2,3-*c*]pyrazole

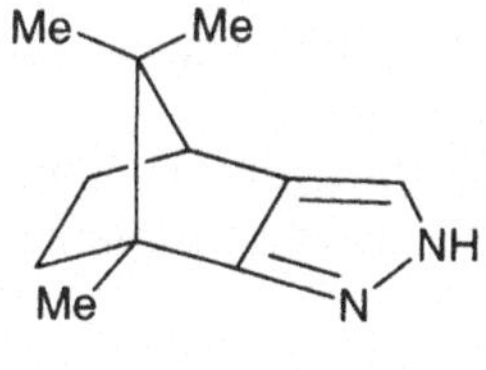

[145], the asymmetric unit contains six independent molecules, comprising two trimers constructed via N-H···N hydrogen bonds.

These interactions are shorter and more linear than generally found for hydrogen bonds of the type N-H···N(sp^2) in organic crystals (as assessed from a survey of known crystal structure in which the hydrogen bond acceptor nitrogen atom is bonded to nitrogen and carbon atoms, as in the pyrazole moiety [146]). For one of the trimers, the conformation of the central six-membered rings (excluding hydrogen atoms) may be described in terms of a distorted half-chair towards an envelope conformation, whereas the other trimer has a slightly distorted 1,3-diplanar conformation. The crystal is built of sheets of alternating trimers. The isotropic ^{13}C chemical shifts for the carbons adjacent to the two nitrogens are consistent with 2*H* tautomers (i.e. all six molecules have the proton in position 2), in agreement with the crystal structure.

The complex hydrogen bonding arrangement in the biomedically important molecule bilirubin IXα, an unsymmetrically substituted tetrapyrrole dicarboxylic acid (shown below), and its dimethyl ester have been probed by using ^{1}H DQ MAS NMR [23].

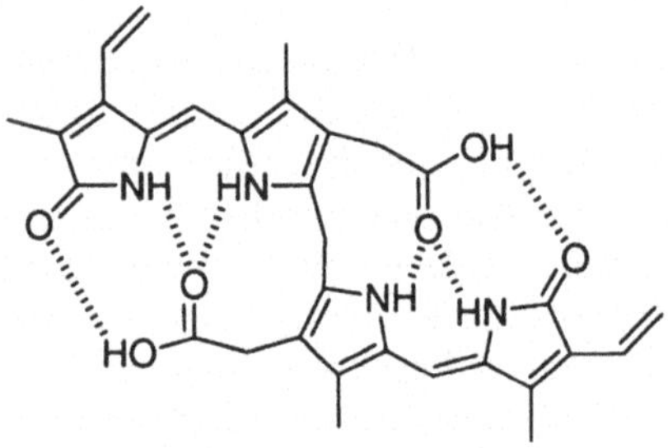

Single crystal X-ray diffraction studies [147] showed that the crystal structure of bilirubin contains multiple hydrogen bonds, as shown above. Employing fast MAS and a high magnetic field, three high-frequency peaks corresponding to the different hydrogen bonded protons were resolved in a ^{1}H MAS NMR spectrum. These resonances were assigned on the basis of the proton-proton proximities identified from a rotor-synchronised ^{1}H DQ MAS NMR spectrum. Analysis of ^{1}H DQ MAS spinning sideband patterns for the NH protons allowed ^{1}H···^{1}H distances to be determined quantitatively. In particular, the distance between the lactam NH and pyrrole NH protons was determined to be 1.86±0.02 Å and the distance between the lactam NH and carboxylic acid OH protons was determined to be 2.30±0.08 Å. In addition, comparison of ^{1}H DQ MAS spinning sideband patterns for bilirubin and its dimethyl ester revealed a significantly longer distance between the two NH protons in the latter case. This study demonstrates the significant opportunities provided by ^{1}H DQ MAS NMR for detailed structural studies of hydrogen bonding arrangements.

Finally, the N···H distance in the hydrogen bonding arrangement adopted by a pair of methyl-substituted benzoxazine dimers has been determined by solid state NMR to be 1.94 Å [148].

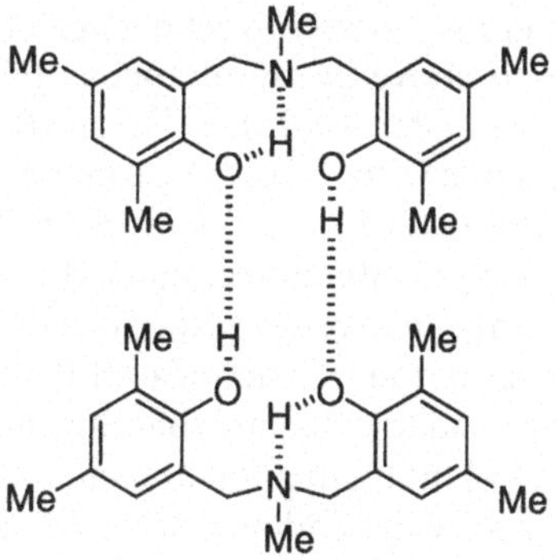

The results indicate that the proton is shared between the nitrogen and oxygen atoms, with a preference for O-H rather than N-H character. Further advanced solid state NMR methods were used to measure the N···H distance via the N···H dipolar coupling.

4.2
Dynamic Aspects of Hydrogen Bonding Arrangements in Solids

4.2.1
Carboxylic Acids

Carboxylic acids are known to form hydrogen bonded dimers in the gas phase, as well as in the liquid and solid phases. There are two ways of forming the hydrogen bonded dimers, as shown below.

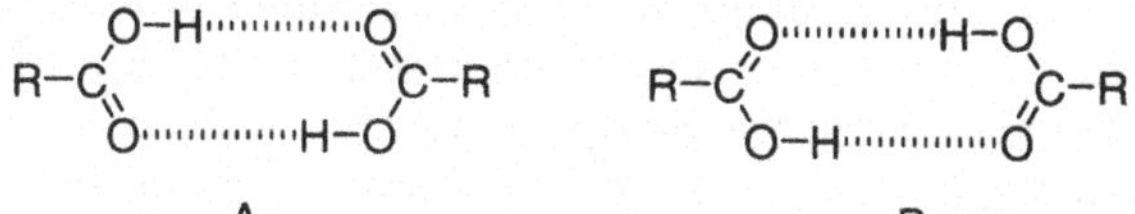

A B

In the gas phase, the two configurations (A and B) are degenerate and the potential energy curve for proton transfer has two minima of equal depth. However, this degeneracy can be removed in the solid state by the effects of the crystal environment [149]. In this regard, benzoic acid and its derivatives have been studied in detail using various techniques in the solid state. First, we note that X-ray diffraction [149] and IR spectroscopy [150] studies have established that there is disorder between configurations A and B for many benzoic acids in the solid state. This disorder may be either dynamic [151] or static [149], and detailed solid state NMR investigations have been undertaken by several groups to explore this issue.

The crystal structure of benzoic acid is monoclinic and contains four molecules per unit cell in the form of two magnetically inequivalent dimers with equal values of the chemical shift tensors but with their principal axis systems oriented differently [152]. Each dimer may interconvert between the two hydrogen bonding configurations discussed above by simultaneous proton transfer along the two hydrogen bonds. Single crystal solid state ^{13}C NMR measurements on benzoic acid (C_6H_5COOH) enriched with ^{13}C at the carboxylic acid carbon have been carried out to determine the rate of proton transfer in the benzoic acid dimer [153]. It was found that the rate of interconversion between configurations A and B is sufficiently rapid that only an average resonance line is observed. From the observed angular dependence of the ^{13}C chemical shift, the chemical shift tensors were determined, and were used to calculate that the energy difference between the two configurations is 0.4 kJ mol^{-1}, in good agreement with the value obtained previously from the temperature dependence of the IR spectrum. From ^{1}H NMR spin-lattice relaxation time (T_1) measurements, the barrier for interconversion between the two configurations was calculated to be (4.9±0.08) kJ mol^{-1}. Consideration of the results of T_1 measurements for C_6H_5COOH and C_6D_5COOH was used to verify that proton transfer along the hydrogen bonds is responsible for the proton relaxation. Below 120 K, plots of $\ln(T_1)$ vs reciprocal temperature deviate from the theoretical curve, and it was suggested that this is due to proton transfer occurring via a tunnelling mechanism. The tunnelling mechanism has been the subject of further detailed NMR studies involving ^{2}H NMR T_1 measurements of benzoic acid [154] and *m*-iodobenzoic acid, 2,3-dimethoxybenzoic acid and Feist's acid [155], and involving ^{1}H NMR T_1 and incoherent quasielastic neutron scattering (IQNS) measurements of diglycolic acid, suberic acid, benzoic acid, terephthalic acid and malonic acid [156] and dodecanoic acid [157].

In a subsequent paper, Nagaoka et al. [158] extended their studies to include decanoic acid and other monosubstituted derivatives of benzoic acid using both ^{1}H NMR T_1 data and IR measurements. The main results of this work were: (i) that the proton transfer processes in benzoic acid, *m*- and *p*-substituted derivatives of benzoic acid and decanoic acid have low values of activation energy in the range 4.9–6.0 kJ mol^{-1}, consistent with results from ab initio calculations [159], and (ii) that the proton transfer processes in *o*-chloro- and *o*-bromobenzoic acids have much higher values of activation energy in the range 54–59 kJ mol^{-1}, although the underlying reasons for this difference were not established (see, however, the discussion below).

p-Toluic acid has been the subject of detailed NMR studies by Ernst et al. [160]. The crystal structure [161] of *p*-toluic acid contains hydrogen bonded dimers, with disorder in the hydrogen bonding evident from the fact that the two C-O bond lengths are almost equal. To investigate this disorder, solid state ^{1}H NMR studies of *p*-toluic acid-d_7 were carried out. From the temperature dependence of the ^{1}H NMR dipolar coupling tensor and ^{1}H spin-lattice relaxation times, the dynamic character of the disorder was deduced and the process was assigned as a correlated double proton transfer mechanism. The potential energy curve for this process is asymmetric due to the effects of the crystal environment. The activation energy for the proton transfer process was estimated to be 4.8 kJ mol^{-1}, with a free energy difference of 1.0 kJ mol^{-1} between the two tautomeric forms. The temperature dependence of the ^{1}H NMR T_1 in terephthalic acid was also reported. The high temperature relaxation is compatible with a classical barrier height of 2.6 kJ mol^{-1} and the low temperature relaxation leads to an apparent activation energy of 0.8 kJ mol^{-1}, which is attributed to a tunnelling mechanism and confirmed via ^{1}H NMR T_1 measurements at lower magnetic fields.

A critical assessment of some of the results reported by Nagaoka et al. [158] and Ernst et al. [160] was published by Furic [162]. The main argument put forward by Furic was that the quoted experimental studies should not be considered as direct evidence for a dynamic double proton exchange in solid carboxylic acids. Instead, it was suggested that interconversion of configurations A and B by means of a 180° rotation of the entire hydrogen bonded eight-membered ring (i.e. the -$CO_2H\cdots HO_2C$- unit) can also explain the observed temperature dependence of NMR parameters. In their reply, Ernst et al. [163] noted that the two mechanisms (shown below) do lead to indistinguishable final states for the NMR observer.

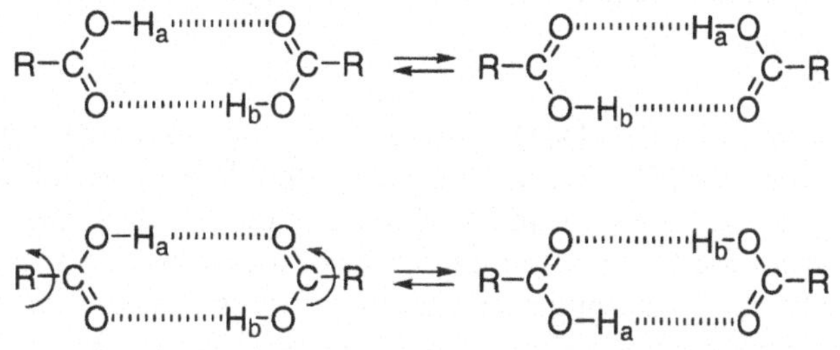

In order to confirm the proton transfer mechanism proposed previously [160], the results of IQNS on terephthalic acid were reported [164]. The jump distance is calculated to be 0.7 Å for the proton transfer model and 2.1 Å for the 180° rotation model – the latter process was ruled out on the basis of the experimental IQNS results, leading to the conclusion that the mechanism of the proton dynamics is indeed a double proton exchange. IQNS results for terephthalic acid and acetylene dicarboxylic acid have also been reported [165]. For both samples, the jump distance was found to be less than 1 Å. For acetylene dicarboxylic acid, single crystal measurements yielded a jump distance of 0.73 Å. The Q-dependence was found to be in excellent agreement with the 2-site jump model. From these results, the 180° rotation model can be ruled out in favour of the proton transfer model.

In their reply to Furic's criticism, Nagaoka et al. suggested [166] that: (i) the 180° rotation model proposed by Furic [162] would have an activation energy much higher than 5 kJ mol^{-1}, and (ii) if rotation of the -$CO_2H\cdots HO_2C$- unit was the mechanism of proton relaxation, the T_1 vs reciprocal temperature curve should be the symmetric curve predicted by classical relaxation theory (i.e. without proton tunnelling effects at low temperature).

Atom-atom potential calculations for carboxylic acid dimers [167] suggested that, while the activation energy is lower for the double proton transfer, both mechanisms can be energetically plausible depending on the structure of the system under investigation.

A third possible dynamic model, which combines features of the two models described above, has been proposed by Haeberlen et al. [168] on the basis of multinuclear solid state NMR studies of dimethylmalonic acid [$Me_2C(COOH)_2$; DMMA]. The ^1H and ^{13}C chemical shift tensors and the ^2H quadrupole interaction tensor (for the carboxylic acid deuteron) were measured for single crystals of DMMA. At room temperature, only half the number of carboxylic acid ^1H and ^2H resonances predicted by symmetry are actually observed. This was attributed to a novel hydrogen exchange process comprising a flip of the whole dimeric unit followed by a rapid concerted jump of the protons along the strongly asymmetric hydrogen bonds. As shown below, the net result of this new dynamic model is a simple mutual exchange of H_a and H_b:

It was suggested that the DMMA dimers have an asymmetric single well potential, rather than asymmetric or symmetric double well potentials. The activation energy derived from lineshape analysis of the ^2H NMR spectra was 66 kJ mol^{-1}, which is similar to the values reported for *o*-chloro- and *o*-bromobenzoic acids. On this basis, it was suggested that high values of activation energies are associated with this mutual hydrogen exchange mechanism, rather than the proton transfer model that occurs for those materials associated with low activation energies. Subsequently [169], ^{17}O NMR studies were undertaken in order to further distinguish between the mutual hydrogen exchange and proton transfer mechanisms for DMMA. The main difference between the two models is that the proton transfer mechanism affects only the ^{17}O-^1H dipole-dipole splitting, whereas the mutual hydrogen exchange mechanism affects the ^{17}O quadrupole splittings of the oxygen atoms of the -$CO_2H\cdots HO_2C$- unit. On the basis of detailed variable-temperature ^{17}O NMR studies of an ^{17}O enriched single crystal of DMMA, it was shown that only the latter model is consistent with the observed spectral changes.

It is interesting to note that, for malonic acid (which is structurally related to DMMA), the activation energy measured from ^1H NMR T_1 measurements [170] is 5.6 kJ mol^{-1}, which is significantly lower than in DMMA and is assigned to proton jumps between the two minima of an asymmetric double well potential. This emphasises the importance of the effect of the crystal packing on the asymmetry of the potential function, which defines the mechanism of the proton dynamics in carboxylic acid dimers.

Finally, high-resolution ^1H NMR techniques employing fast MAS have been used to study the structure and dynamics of a hexabenzocoronene carboxylic acid derivative [25] shown below.

The presence of hydrogen bonded carboxylic acid dimers in the solid was demonstrated from the ^1H DQ MAS NMR spectrum, with the $^1H\cdots{}^1H$ distance determined to be 2.79±0.9 Å. The spectral changes as a function of temperature

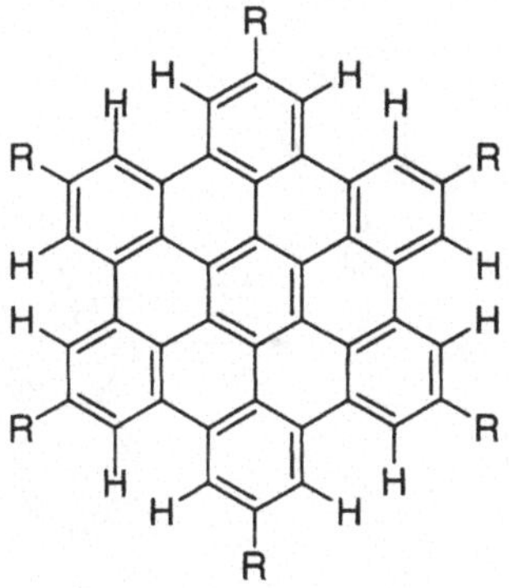

R=$C_{10}H_{20}COOH$

were interpreted in terms of a simple exchange process involving the making and breaking of hydrogen bonds:

$$RCOOH\cdots HOOCR \rightleftarrows 2\,RCOOH$$

The kinetics of the dimer opening transformation were determined, with the activation energy estimated to be 89 kJ mol^{-1}.

4.2.2
Tropolone

Tropolone is known to undergo a tautomeric hydrogen shift shown below. In solution state ^{1}H and ^{13}C NMR spectra, averaged signals due to interconversion between the two tautomeric forms are observed.

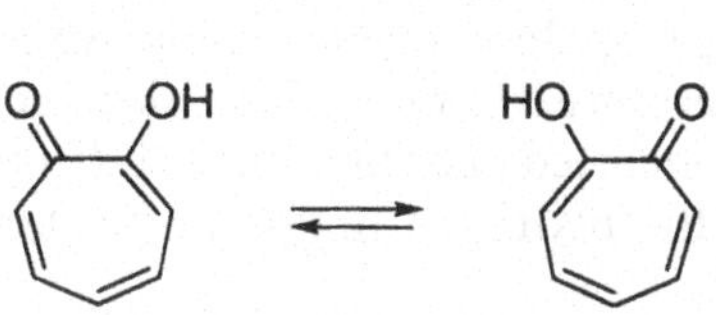

In the crystal structure of tropolone, the molecules are arranged as centrosymmetric hydrogen bonded dimers. The crystals are highly ordered, with the molecules forming coplanar hydrogen bonded pairs. No evidence was found for disorder in the positions of the hydrogens atoms. Each hydroxyl group participates in a bifurcated hydrogen bond with two carbonyl oxygen atoms, one in the same molecule and one in the other molecule of the dimer. It therefore came as surprise when Szeverenyi et al discovered by 2D-exchange ^{13}C MAS NMR [171] that tautomeric hydrogen shifts between hydroxyl and carbonyl oxygen atoms takes place in crystalline tropolone. Obviously, such a process will lead to hydrogen disorder. A dynamic model was suggested, according to which the hydrogen shift proceeds in a concerted manner with a 180° flip of the entire molecule (or dimer), which restores the original orientation of the tropolone

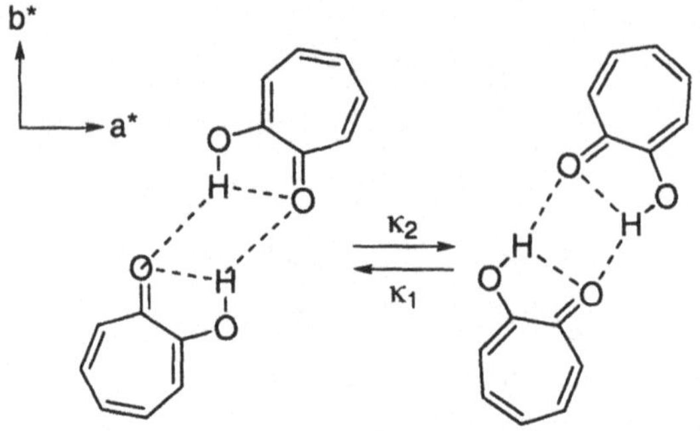

Fig. 8 Projection of the majority (*left*) and minority (*right*) species of the tropolone dimer along the crystal axis c. The rate constants of the exchange of the two species are denoted κ_1 and κ_2

molecules in the crystal structure. The proposed mechanism is also consistent with the high activation energy (109 kJ mol^{-1}) of the dynamic process. On the basis of other experimental results for tropolone (mainly ^{13}C NMR), it was proved that the hydrogen shift is a secondary process that occurs after the carbonyl and hydroxyl oxygen positions become interchanged (by the 180° flip motion) [172–174]. The occurrence of another dynamic process consisting of rapid concerted hydrogen shifts within the dimer was recently suggested on the basis of the orientation dependence and temperature dependence of ^{2}H NMR lineshape and ^{2}H spin-lattice relaxation time measurements for the hydroxyl deuterons in a single crystal of tropolone-d_1 [175]. The results were interpreted in terms of a dynamic hydrogen disorder model in which the hydrogen nuclei move in an asymmetric double well potential. According to this model, the hydrogen bonded dimer structure, as determined by X-ray diffraction, constitutes a majority species in the tropolone crystal, comprising more than 98% of the molecules at room temperature. However, there also exists a tautomeric minority species formed by a concerted back and forth shifting of the hydroxyl hydrogens (deuterons) along the hydrogen bonds to the nearby carbonyl oxygens (Fig. 8).

In principle, the hydrogen shift within the dimer could occur via an intramolecular pathway or an intermolecular pathway, which cannot be distinguished by NMR. The hydrogen shift process between the majority and the minority species results in a modulation of the ^{2}H EFG tensor, thus providing an efficient relaxation mechanism. The concentration of the minority species is too low and its lifetime is too short to make its direct observation possible. Structural information about this species and kinetic and thermodynamic parameters relating to the hydrogen shift process were derived by fitting the measured T_1 values to the dynamic model described above. Interesting comparisons have also been drawn [175] between the hydrogen dynamics in the tropolone dimer and in carboxylic acid dimers.

4.2.3
Alcohols

One of the most revealing applications of single crystal ^2H NMR has been reported by Haeberlen et al. [176]. Advantages provided by this technique were used to study the dynamic properties of the host structure in the clathrate of Dianin's compound with ethanol as the guest molecule.

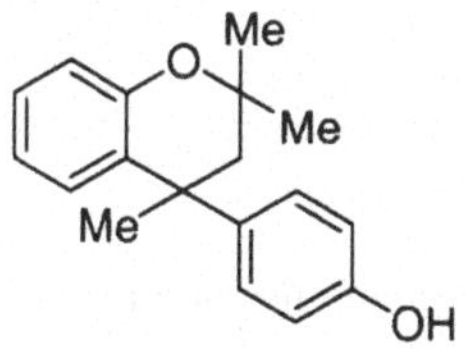

The hydroxyl groups of both the host structure and the guest were deuterated. The host molecules in these solid inclusion compounds form cages in which the guest molecules are trapped, and the ends of the cages are formed by hexagons of oxygen atoms of the hydroxyl groups linked by hydrogen bonds. From the temperature dependence of the ^2H NMR spectra, it was suggested that the hydroxyl deuterons of the host jump between two unequally populated sites via "approximate" rotation of each hydroxyl groups about its C-O bond (Fig. 9).

The activation energy for this dynamic process was estimated to be 33.1 kJ mol^{-1}. The term "approximate" rotations was used since the C-O-D bond angles for the major and the minor sites are slightly different (112.5° and 116.0° respectively) and the motion is therefore also associated with a slight change in the molecular geometry. Independent rotations of the hydroxyl groups was ruled out by the absence of dipolar fine structure in the single crystal ^2H NMR spectra, and it was suggested that the six hydroxyl groups jump in a concerted manner. Interestingly, the fractional populations of the major and minor sites were found to be temperature dependent.

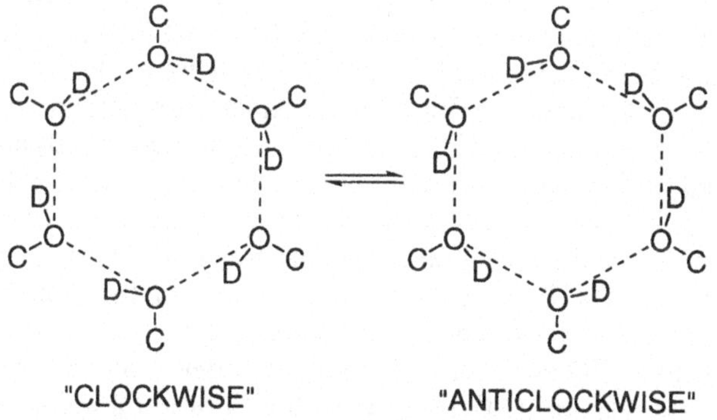

Fig. 9 Schematic representation of the two hydrogen bonding arrangements involved in concerted rotation of the hydroxyl groups about the C-O bonds in the clathrate of Dianin's compound [176]

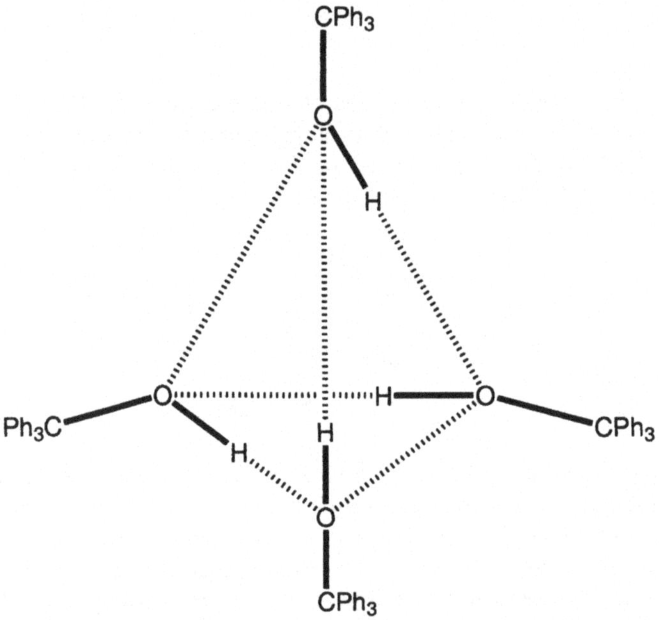

Fig. 10 Schematic representation of the tetramer in the crystal structure of triphenylmethanol [179]. The hydrogen bonding arrangement shown is only one of several possible hydrogen bonding arrangements for the tetramer

Dynamic properties of the hydrogen bonding arrangement in a selectively deuterated sample of solid triphenylmethanol (Ph_3COD) have been studied using solid state ^{2}H NMR [177, 178]. In the crystal structure (Fig. 10), the molecules form hydrogen bonded tetramers, with the oxygen atoms positioned approximately at the corners of a tetrahedron [179].

The tetramer has point symmetry C_3 (rather than T_d); three of the Ph_3COD molecules (denoted as "basal") are related to each other by a threefold rotation axis, and the fourth molecule (denoted as "apical") lies on this axis. Thus, the oxygen atoms from the four molecules in the tetramer form a pyramidal arrangement with an equilateral triangular base, and the O···O distances are consistent with the tetramer being held together by O-H···O hydrogen bonds. The ^{2}H NMR lineshape for Ph_3COD varies as a function of temperature, demonstrating that the hydrogen bonding arrangement is dynamic. Several plausible dynamic models were considered, and it was found that only one model gives a good fit to the experimental ^{2}H NMR spectra across the full temperature range studied. In this model, the deuteron of the apical molecule undergoes a 3-site 120° jump motion by rotation about the C-O bond with equal populations of the three sites, whereas the deuterons of the basal molecules undergo a 2-site 120° jump motion, by rotation about their C-O bonds. In addition, each deuteron undergoes rapid libration about the relevant C-O bond with the libration amplitude increasing as a function of temperature. The behaviour of the basal molecules was interpreted in terms of the existence of two possible hydrogen bonding arrangements (described as "clockwise" and "anticlockwise") on the basal plane of the pyramid [177]:

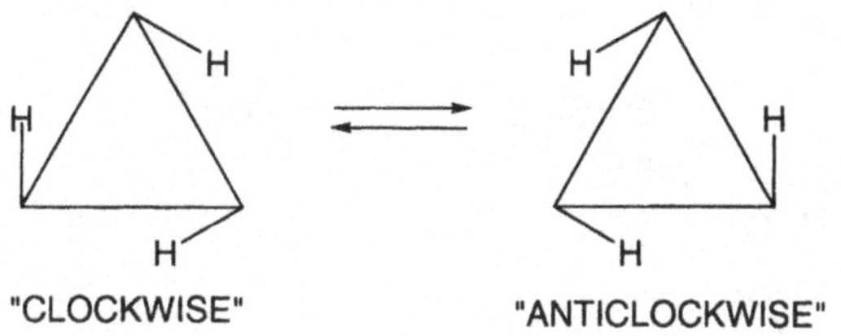

The 2-site 120° jump motion for the basal molecules "switches" between these two hydrogen bonding arrangements and clearly requires correlated jumps of the hydroxyl groups of all three basal molecules. On the assumption of Arrhenius behaviour for the temperature dependence of the jump frequencies, the activation energies for the jump motions of the apical and basal deuterons were estimated to be 10 and 21 kJ mol^{-1}, respectively. This dynamic model was further supported by analysis of the dependence of the quadrupole echo ^2H NMR lineshape on the echo delay and consideration of ^2H NMR spin-lattice relaxation time data.

Similarly, the dynamic properties of the hydroxyl groups in a selectively deuterated sample of triphenylsilanol (Ph_3SiOD) have been studied [180]. The crystal structure of triphenylsilanol is different from that of triphenylmethanol and contains eight crystallographically independent molecules, which are arranged in two tetrameric building units. Within each of these tetrameric units, the four silicon atoms are arranged in the form of a slightly distorted square, with the oxygen atoms of the four hydroxyl groups involved in O-H···O hydrogen bonding. All eight crystallographically inequivalent Si sites are resolved in the ^{29}Si CP MAS NMR spectrum within the chemical shift range −11 ppm to −16 ppm (Fig. 11).

The temperature dependence of the quadrupole echo ^2H NMR lineshape and ^2H NMR spin-lattice relaxation time measurements demonstrated that the hydrogen bonding arrangement is dynamic.

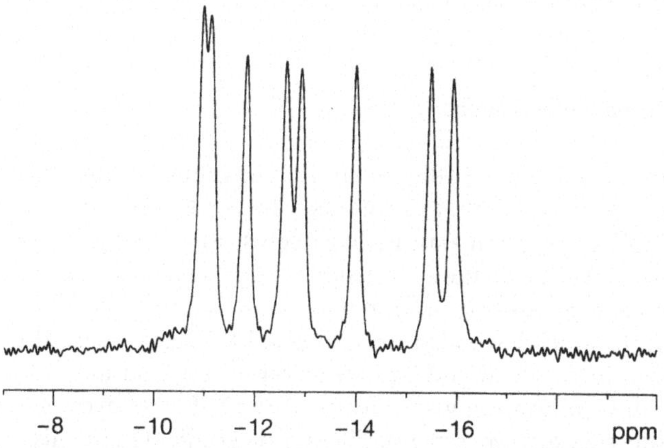

Fig. 11 Solid state ^{29}Si CP MAS NMR spectrum of Ph_3SiOH recorded at 363 K [180]

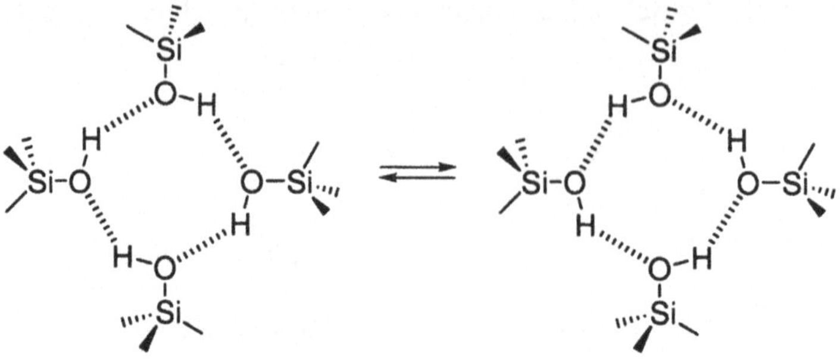

The dynamic process is interpreted as interconversion between "clockwise" and "anticlockwise" hydrogen bonding arrangements within each tetrameric unit, via a 2-site jump motion of each hydroxyl deuteron about the Si-OD bond. The activation energy for the dynamic process was estimated to be 35 kJ mol^{-1}.

In another study, ^1H NMR has been applied to investigate proton dynamics in anhydrous α-D-glucose [181], in which all five hydrogen bond donors in each molecule form intermolecular hydrogen bonds. The structure is layered. Molecules within a layer interact via the shortest hydrogen bond, with weaker hydrogen bonds linking adjacent layers. The CH_2OH group exists in both *gauche* and *trans* forms within the structure. At low temperature, the *trans* rotamer is much less populated than the *gauche* rotamer. The ^1H NMR relaxation times T_1 and $T_{1\rho}$ were found to be relatively long, suggesting that the relaxation mechanism is weak. The observation of minima in the relaxation times as a function of temperature proved that the dipolar interaction is modulated by thermally activated molecular motions. It was shown that the *trans-gauche* rearrangement of the CH_2OH group and the jump motion of an OH group proton between two equilibrium sites in a hydrogen bond are the motions contributing to the observed ^1H NMR relaxation times T_1 and $T_{1\rho}$.

4.2.4
Amino Acids, Peptides and Proteins

Crystalline amino acids have often been used as model compounds for probing functional group interactions in proteins. The 3-site 120° jump motion of the ammonium ($-NH_3^+$) group in alanine has been studied using ^2H NMR lineshape analysis and by considering the anisotropy of the ^2H spin-lattice relaxation [182]. The activation energy for this motion was estimated to be 40.5 kJ mol^{-1}.

^2H NMR techniques have also been applied to characterise the ammonium group reorientation in the α and β polymorphs of L-glutamic acid [183]. In both polymorphs, the ammonium group forms three N-H···O hydrogen bonds, with only small differences (from neutron diffraction studies) in the distances and angles that define the hydrogen bonding geometries. In spite of these small differences in geometry, however, significant differences in the rate of ammonium group reorientation are observed (at a given temperature) in the two polymorphs. From ^2H NMR lineshape analysis, the activation energy for the reorientation of the $-ND_3^+$ group was determined to be 47 kJ mol^{-1} for the α phase and 34 kJ mol^{-1} for the β phase, in good agreement with results from ^2H NMR spin-lattice relaxation time data (48 kJ mol^{-1} for the α phase and 39 kJ mol^{-1} for the β phase). The small differences in hydrogen bonding geometries involving the $-NH_3^+$ group in the α and β phases suggest that the hydrogen bonding is stronger in the α phase, consistent with the observation of a higher activation energy for the ammonium group reorientation in this polymorph.

Hydrogen bonding effects on ammonium group rotation rates have also been studied in other crystalline amino acids [184]. ^1H spin-lattice relaxation times and ^2H NMR lineshapes were measured for D-, D,L- and L-aspartic acid, two polymorphs of glycine, alanine and leucine in the temperature range from 233 to 383 K. The activation energies for ammonium group rotation were determined to be 27 kJmol^{-1} for D- or L-aspartic acid, 22 kJ mol^{-1} for D,L-aspartic acid, 24 and 30 kJ mol^{-1} for the α and γ forms of glycine respectively, 40 kJ mol^{-1} for L-alanine and 49 kJ mol^{-1} for L-leucine. Differences in the hydrogen bonding environments around the $-NH_3^+$ groups were proposed as a basis for the different activation energies observed.

4.2.5
Urea, Thiourea and Their Inclusion Compounds

An example of the application of ^2H NMR to probe dynamics of hydrogen bonded solids concerns the study of dynamics of crystalline urea and urea inclusion compounds containing alkane [i.e. $Me(CH_2)_nMe$/urea-d_4] and α,ω-dibromoalkane [i.e. $Br(CH_2)_nBr$/urea-d_4] guest molecules. In these inclusion compounds [185, 186], the urea molecules form an extensively hydrogen bonded host structure containing parallel one-dimensional tunnels that are densely packed with the guest molecules. The dynamic properties of the urea molecules in the nonadecane/urea-d_4 inclusion compound have been studied by powder [126] and single crystal [187] ^2H NMR leading to the proposal that the urea molecules undergo 180 ° jumps about their C=O axes, with no evidence (on the ^2H NMR timescale) for rotation of the NH_2 groups about the C-N bonds. To probe whether the exact nature of the guest molecules (and particularly the presence of different types of functional group on the guest molecules) could have a significant bearing upon the urea jump motion, the urea dynamics in the $Br(CH_2)_nBr$/urea inclusion compounds were also studied [188]. Again, the ^2H NMR lineshapes can be simulated successfully on the basis of a 2-site 180° jump motion about the C=O axis of the urea molecule. Qualitative features of the ^2H NMR spectra are identical for all the $Br(CH_2)_nBr$/urea-d_4 inclusion compounds studied. The spectra recorded at 293 K for the urea-d_4 inclusion compounds with $Br(CH_2)_8Br$, $Br(CH_2)_9Br$ and $Br(CH_2)_{10}Br$ guest molecules were fitted well by a spectrum simulated using jump frequency $\kappa=4\times10^6$ s^{-1}, whereas for the $Br(CH_2)_7Br$/urea-d_4 inclusion compound, the best fit value of κ at 293 K is 1.5×10^6 s^{-1}.

A 2-site 180° jump motion of the urea molecule about its C=O axis is also believed to occur in the pure crystalline phase of urea [189–192] above ambient temperature, and it has been proposed that simultaneous rotation about the C-N bond may also occur [189, 190, 193, 194].

Investigations of the molecular motion in thiourea-d_4 have been undertaken by various groups. The structure of the high-temperature paraelectric phase of thiourea-d_4 has been determined previously by diffraction methods [195, 196]; the orthorhombic structure has four molecules in the unit cell arranged on planes with alternating molecular orientation. Each molecule interacts with two of its neighbours through four hydrogen bonds forming a hydrogen bonded network. The structure of the approximately planar thiourea-d_4 molecule is shown below.

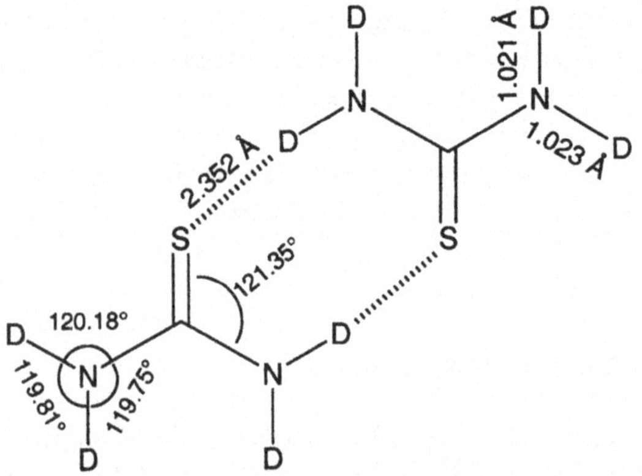

^{2}H NMR lineshape analysis based on automated non-linear least squares fitting (Fig. 12) was used to establish that a 2-site 180° jump motion occurs about the C=S bond, together with small angle librational motion [31]. The activation energy for the 2-site 180° jump motion about the C=S bond was estimated to be 47.8 kJ mol^{-1}, in good agreement with the value (46.4 kJ mol^{-1}) obtained by variable temperature ^{2}H MAS NMR [197]. The MAS experiment was also used to characterise the dynamics of the slow C-N rotation in thiourea-d_4, for which the activation energy was reported to be 56.3 kJ mol^{-1} [197].

4.2.6
Pyrazoles, Imidazoles and Triazoles

Detailed studies of proton disorder in 3,5-dimethylpyrazole have been undertaken by Elguero and co-workers [198]. Annular tautomerism is defined as prototropy involving exclusively ring nitrogens and is common in all *N*-unsubstituted azoles. High-resolution solid state ^{13}C NMR studies of azoles have revealed two general features: (i) "narrow" singlets corresponding to a unique tautomer are usually observed, and (ii) the structure of the tautomer is in agreement with that established from X-ray diffraction data [199, 200]. However, for 3,5-dimethylpyrazole, the high-resolution solid state ^{13}C NMR spectrum recorded at

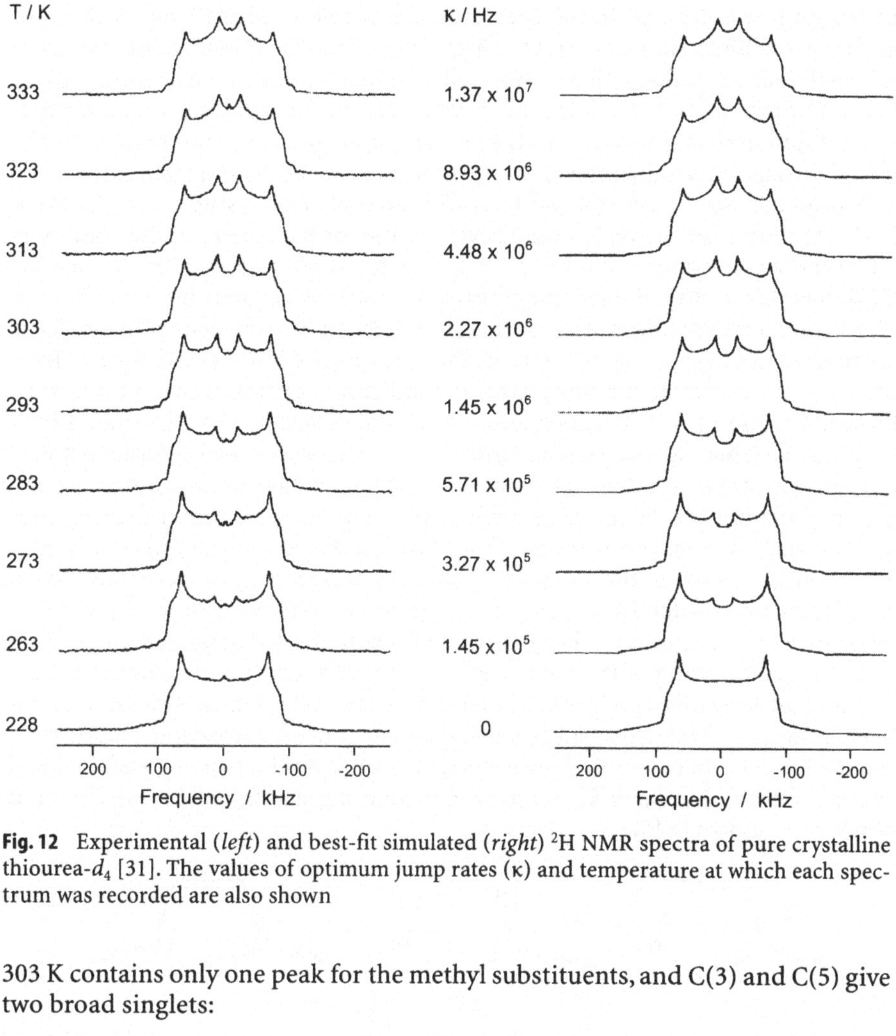

Fig. 12 Experimental (*left*) and best-fit simulated (*right*) ^{2}H NMR spectra of pure crystalline thiourea-d_4 [31]. The values of optimum jump rates (κ) and temperature at which each spectrum was recorded are also shown

303 K contains only one peak for the methyl substituents, and C(3) and C(5) give two broad singlets:

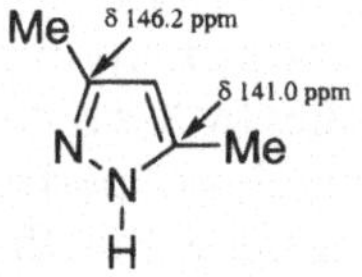

At low temperature there are two resolved peaks for the methyl substituents, but the observed splitting is reduced on increasing temperature. X-ray diffraction results show that the unit cell consists of a cyclic trimeric arrangement to 3,5-dimethylpyrazole molecules. Within this trimer, the 3,5-dimethylpyrazole molecules have C_{2v} symmetry, the cyclic trimer has threefold symmetry, and the hydrogen involved in the tautomeric process is refined with half occupancy. These results indicate that, at room temperature, a trimer-trimer intermolecular tautomerism takes place in 3,5-dimethylpyrazole. In contrast, for pyrazole, the N-H hydrogen is well located in the crystal structures determined from both X-ray and neutron diffraction data, and the high resolution ^{13}C CP MAS NMR spectrum has well defined peaks with no observed broadening due to a dynamic process [199]. The explanation given for the lack of dynamic behaviour is that intermolecular tautomerism is less favourable because of the packing arrangement (a distorted tetrameric arrangement) of the molecules in the crystal structure.

Proton disorder in several solid pyrazoles has also been studied by ^{15}N NMR [201]. As discussed above, 3,5-dimethylpyrazole forms trimers in the solid state and undergoes a concerted triple proton transfer. High-resolution solid state ^{15}N NMR has shown that, at low temperature, there are two signals for the nitrogens of 3,5-dimethylpyrazole indicating that protonated and nonprotonated nitrogens are present in equal concentrations. As the temperature is increased, the two lines broaden and coalesce into one sharp line indicating proton transfer with equilibrium constant $K \approx 1$. This behaviour is also observed for 3,4-diphenyl-4-bromopyrazole which forms a cyclic dimer in the solid state and 3,5-diphenylpyrazole which forms a cyclic tetramer. The observed tautomeric processes are assigned to multiple proton transfer. It is interesting to note that the rate of proton transfer (as deduced from the ^{15}N NMR lineshape analysis) first decreases and then increases as the number of protons transferred is increased, which could indicate a switch from a concerted process to a stepwise process [202] (the latter may be expected for a large cyclic hydrogen bonded arrangement).

Hydrogen bonding of the type N-H···N formed between molecules of imidazole and its derivatives is closely related to a variety of biological systems and has been a subject of extensive studies using a variety of spectroscopic and diffraction techniques. In crystalline imidazole, the molecules form a one-dimensional chain of intermolecular N-H···N hydrogen bonding, a schematic representation of which is shown below.

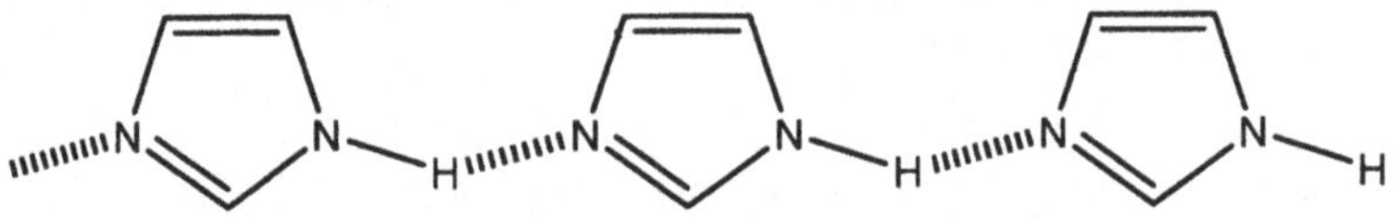

On the basis of electronic conductivity measurement [203] and ^{1}H NMR results [204] it was postulated that the protons migrate through this intermolecular chain. However, structural studies by X-ray and neutron diffraction [205, 206] indicated that the hydrogen atom is almost perfectly localised and does not show any evidence of intermolecular transfer within the hydrogen bond. One- and two-dimensional ^{15}N exchange CP MAS NMR techniques as well as static ^{15}N NMR have been applied recently to study the possibility of proton transfer in imidazole [207]. In the 2D EXSY spectrum, cross peaks were observed between the main ^{15}N resonance peaks for -N= and -N<, implying that magnetization exchange takes place between the -N= and -N< environments. Based on the dependence of the exchange rate on the power of the ^{1}H decoupling field, it was concluded that the magnetization transfer in crystalline imidazole is dominated by a proton-driven spin-diffusion mechanism, rather than by a chemical exchange mechanism.

An interesting interpretation of temperature dependent lineshapes based on population changes as a function of temperature has been proposed recently [208]. In particular, variable temperature ^{15}N CP MAS NMR has been employed to investigate proton transfer dynamics and N-H bond lengthening in N-H···N hydrogen bonds.

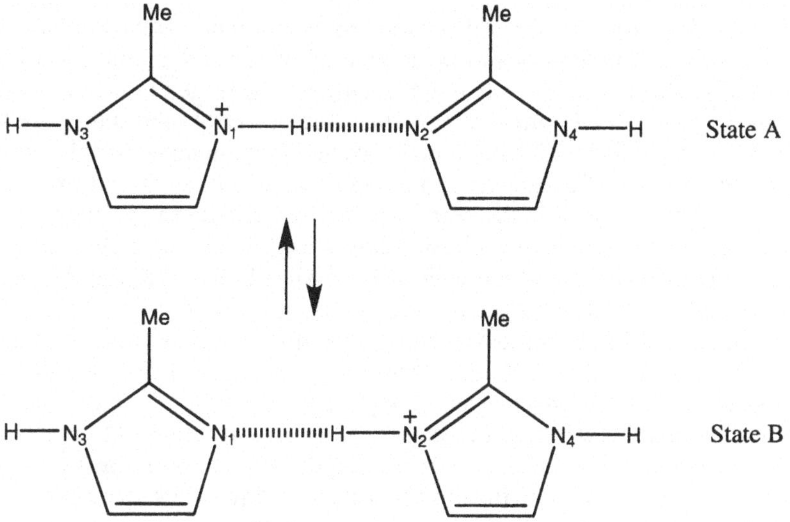

Hydrogen bonded pairs of the 2-methylimidazolium cation and 2-methylimidazole were used as a model for enzyme active sites. For the chloride and bromide salts, temperature dependent lineshapes were simulated using the assumption that a rapid (on the NMR timescale) hydrogen transfer is observed at all temperatures in the range 200–280 K, but the populations of the two configurations A and B vary with temperature. This assumption was supported by the fact that there is no significant broadening and no exchange cross-peaks are observed in 2D exchange ^{15}N NMR spectra. It was found that the populations of configurations A and B at room temperature are equal, whereas at 200 K the proton of the N-H···N hydrogen bond is effectively trapped in one of the configurations.

Finally, triazoles ([$^{15}N_2$]-labelled 3,5-dibromo-1*H*-1,2,4-triazole and 3,5-dichloro-1*H*-1,2,4-triazole) have also been studied by ^{15}N CP MAS NMR [209]. These compounds form cyclic trimers in their crystal structures.

The ^{15}N CP MAS NMR spectra showed temperature-dependent lineshapes which were analysed in terms of near-degenerate triple proton transfer processes, schematically presented above. The equilibrium constants were found

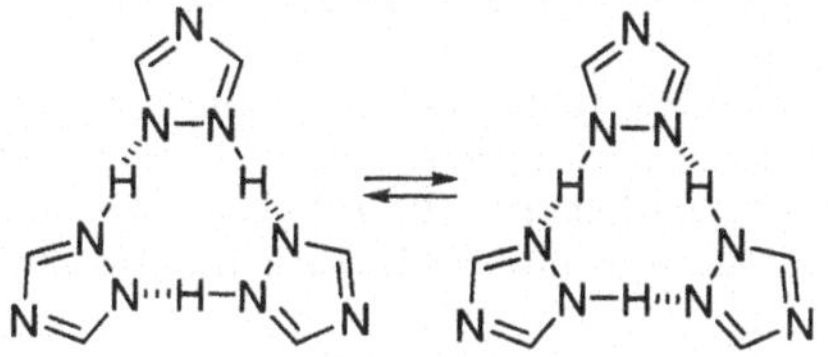

to be slightly different from 1. Rate constants of the triple proton transfer processes were obtained at different temperatures by lineshape analysis.

5 Concluding Remarks

It should be clear from the above discussion that solid state NMR techniques have made significant contributions towards advancing our understanding of a wide range of structural and dynamic issues relating to hydrogen bonding in solids. With the continual development of new and more powerful solid state NMR techniques, we may predict with confidence that even more detailed insights into the fundamental nature of hydrogen bonded systems will be obtained from applications of solid state NMR in the years to come. Nevertheless, it is pertinent to recall that advances in scientific understanding are seldom based on the results obtained from one technique alone. Indeed, for many of the systems and processes discussed in this article, a detailed understanding has arisen by combining the information gained from solid state NMR experiments with the information obtained from the application of other techniques, including diffraction methods, other spectroscopic approaches and computational studies. In many cases, the use of judiciously chosen combinations of different solid state NMR approaches has also been a major factor in understanding the systems of interest. Solid state NMR spectroscopy is a powerful and versatile technique in its own right, but the technique is at its most potent when used in combination with other experimental and computational methods as part of a carefully-planned, multi-technique research strategy.

6 References

1. Abragam A (1961) Principles of nuclear magnetism. Clarendon Press, Oxford
2. Mehring M (1983) Principles of high resolution NMR in solids, 2nd edn. Springer, Berlin Heidelberg New York
3. Harris RK (1983) Nuclear magnetic resonance spectroscopy. Pitman Books, London
4. Fyfe CA (1983) Solid state NMR for chemists. CFC Press, Guelph, Ontario
5. Ernst RR, Bodenhausen G, Wokaun A (1986) Principles of nuclear magnetic resonance in one and two dimensions. Clarendon Press, Oxford
6. Slichter CP (1989) Principles of magnetic resonance, 3rd edn. Springer, Berlin Heidelberg New York
7. Haeberlen U (1976) High resolution NMR in solids – selective averaging. Academic Press, New York
8. Gerstein BC, Dybowski CR (1985) Transient techniques in NMR of solids. Academic Press, Orlando
9. Stejskal EO, Memory JD (1994) High resolution NMR in the solid state. Oxford University Press, New York Oxford
10. Granger P, Harris RK (1990) (eds) Multinuclear magnetic resonance in liquids and solids – chemical applications. Kluwer Academic Publishers, Netherlands
11. Brown SP, Spiess HW (2001) Chem Rev 101:4125
12. Bryce DL, Bernard GM, Gee M, Lumsden MD, Eichele K, Wasylishen RE (2001) Can J Anal Sci Spectrosc 46:46

13. Schnell I, Spiess HW (2001) J Magn Reson 151:153
14. Horsewill AJ, Ikram A (1996) Physica B 226:202
15. Brougham DF, Horsewill AJ, Ibberson RM, Ikram A, McDonald PJ, PinterKrainer M (1996) J Chem Phys 105:979
16. Latanowicz L, Reynhardt EC (2001) Chem Phys Lett 341:561
17. Waugh JS, Huber LM, Haeberlen U (1968) Phys Rev Lett 20:180
18. Gerstein BC, Pemberton RG, Wilson RC, Ryan LM (1977) J Chem Phys 66:361
19. Scheler G, Haubenreisser U, Rosenberger H (1981) J Magn Reson 44:134
20. Maciel GE, Bronnimann CE, Hawkins BL (1990) Adv Magn Reson 14:125
21. Hafner S, Spiess HW (1997) Solid State Nucl Magn Reson 8:17
22. Schnell I, Brown SP, Low HY, Ishida H, Spiess HW (1998) J Am Chem Soc 120:11,784
23. Brown SP, Zhu XX, Saalwächter K, Spiess HW (2001) J Am Chem Soc 123:4275
24. Costa PR, Gross JD, Hong M, Griffin RG (1997) Chem Phys Lett 280:95
25. Brown SP, Schnell I, Brand JD, Mullen K, Spiess HW (2000) Phys Chem Chem Phys 2:1735
26. Schmitt H, Zimmermann H, Korner O, Stumber M, Meinel C, Haeberlen U (2001) J Magn Reson 151:65
27. Hoatson GL, Vold RL (1994) NMR Basic Principles Progr 32:3
28. Spiess HW (1991) Chem Rev 91:1321
29. Wittebort RJ, Olejniczak ET, Griffin RG (1987) J Chem Phys 86:5411
30. Greenfield MS, Ronemus AD, Vold RL, Vold RR, Ellis PD, Raidy TR (1987) J Magn Reson 72:89
31. Aliev AE, Harris KDM (1998) Magn Reson Chem 36:855
32. Ok JH, Vold RR, Vold RL (1989) J Phys Chem 93:7618
33. Aliev AE, Harris KDM, Shannon IJ, Glidewell C, Zakaria CM, Schofield PA (1995) J Phys Chem 99:12,008
34. Torchia DA, Szabo A (1982) J Magn Reson 49:107
35. Morrison C, Bloom M (1993) J Magn Reson Ser A 103:1
36. Schmidt C, Blümich B, Spiess HW (1988) J Magn Reson 79:269
37. Blümich B, Spiess HW (1988) Angew Chem Int Ed Engl 27:1655
38. Schaefer D, Leisen J, Spiess HW (1995) J Magn Reson Ser A 115:60
39. Gerardymontouillout V, Malveau C, Tekely P, Olender Z, Luz Z (1996) J Magn Reson Ser A 123:7
40. Brown MJ, Vold RL, Hoatson GL (1996) Solid State Nucl Magn Reson 6:167
41. Brown MJ, Hoatson GL, Vold RL (1996) J Magn Reson Ser A 122:165
42. Jonsen P (1990) Chem Phys Lett 170:311
43. Williams JC, McDermott AE (1993) J Phys Chem 97:12,393
44. Caravatti P, Bodenhausen G, Ernst RR (1982) Chem Phys Lett 89:363
45. Santos RA, Wind RA, Bronnimann CE (1994) J Magn Reson Ser B 105:183
46. Gu Z, Ridenour CF, Bronniman CE, Iwashita T, McDermott A (1996) J Am Chem Soc 118:822
47. Chuang L-S, Maciel G (1996) J Am Chem Soc 118:401
48. Chuang L-S, Kinney DR, Maciel G (1993) J Am Chem Soc 115:8695
49. Naito A, Nishimura K, Kimura S, Aida M, Yosuoka N, Tuzi S, Saitô H (1996) J Phys Chem 100:14,995
50. Kimura S, Naito A, Saitô H, Ogawa K, Shoji A (2001) J Mol Struct 562:197
51. Shoji A, Ando S, Kuroki S, Ando I, Webb GA (1993) Ann Rep NMR Spectrosc 26:55
52. Goward GR, Schnell I, Brown SP, Spiess HW (2001) Magn Reson Chem 39:S5
53. Fu RQ, Cotten M, Cross TA (2000) J Biomolec NMR 16:261
54. Yamauchi K, Kuroki S, Ando I, Ozaki T, Shoji A (1999) Chem Phys Lett 302:331
55. Wu G, Yamada K, Dong S, Grondey H (2000) J Am Chem Soc 122:4215
56. Dong S, Yamada K, Wu G (2000) Z Naturforsch Teil A 55:21
57. Howes AP, Jenkins R, Smith ME, Crout DHG, Dupree R (2001) Chem Commun 1448
58. Frydman L, Harwood JS (1995) J Am Chem Soc 117:5367
59. Fernandez C, Amoureux JP (1995) Chem Phys Lett 242:449

60. Wu G, Dong S (2001) J Am Chem Soc 123:9119
61. Brunner E, Karge HG, Pfeifer H (1992) Z Phys Chem 176:173
62. Rundle RE, Parasol M (1952) J Chem Phys 20:1487
63. Lord RC, Merrifield RE (1953) J Chem Phys 21:166
64. Blinc R, Hadži D (1966) Nature 212:1307
65. Berglund B, Lindgren J, Tegenfeldt J (1978) J Mol Struct 43:179
66. Hunt MJ, Mackay AL (1974) J Magn Reson 15:402
67. Soda G, Chiba T (1969) J Chem Phys 50:439
68. Hunt MJ, Mackay AL (1976) J Magn Reson 22:295
69. Sternberg U (1988) Mol Phys 63:249
70. Sternberg U, Brunner E (1994) J Magn Reson Ser A 108:142
71. Michal CA, Wehman JC, Jelinski LW (1996) J Magn Reson Ser B 111:31
72. Camus S, Harris KDM, Johnston RL (1997) Chem Phys Lett 276:186
73. Turner GW, Johnston RL, Harris KDM (2000) Chem Phys 256:159
74. Ono S, Taguma T, Kuroki S, Ando I, Kimura H, Yamauchi K (2002) J Mol Struct 602:49
75. Berglund B, Vaughan RW (1980) J Chem Phys 73:2037
76. Jeffrey GA, Yeon Y (1986) Acta Crystallogr B42:410
77. Yesinowski JP, Eckert H (1987) J Am Chem Soc 109:6274
78. Harris RK, Jackson P, Merwin LH, Say BJ, Hagele G (1988) J Chem Soc Faraday Trans 84:3649
79. Rosenberger H, Scheler G, Moskvich YN (1989) Magn Reson Chem 27:50
80. Ditchfield R (1976) J Chem Phys 65:3123
81. Wu G, Freure CJ, Verdurand E (1998) J Am Chem Soc 120:13,187
82. Kaliaperumal R, Sears REJ, Ni QW, Furst JE (1989) J Chem Phys 91:7387
83. Bastow TJ, Elcombe MM, Howard CJ (1986) Solid State Commun 59:257
84. Yamauchi K, Kuroki S, Fuji K, Ando I (2000) Chem Phys Lett 324:435
85. Brunner E, Sternberg U (1998) Prog Nucl Magn Reson 32:21
86. Duncan TM (1990) A compilation of chemical shift anisotropies. Farragut Press, Madison Wisconsin
87. Rohlfing CM, Allen LC, Ditchfield R (1983) J Chem Phys 79:4958
88. Pfrommer BG, Mauri F, Louie SG (2000) J Am Chem Soc 122:123
89. Maciel GE, Traficante DD (1966) J Am Chem Soc 88:220
90. Matiers PM, Jeffrey GA, Ruble JF (1988) Acta Cryst B44:516
91. Etter MC, Hoye RC, Vojta GM (1988) Crystallogr Rev 1:281
92. Etter MC, Reutzel SM, Vojta GM (1990) J Mol Struct 237:165
93. Imashiro F, Maeda S, Takegashi K, Terao T, Saika A (1983) Chem Phys Lett 99:189
94. Takegoshi K, Naito A, McDowell CA (1985) J Magn Reson 65:34
95. Imashiro F (1982) Chem Phys Lett 92:642
96. Imashiro F (1983) Chem Phys Lett 99:189
97. Takeda N, Kuroki S, Kurosu H, Ando I (1999) Biopolymers 50:61
98. Hatzell CJ, Whitfield M, Oas TG, Drobny GP (1987) J Am Chem Soc 109:5966
99. Asakura T, Yamazaki Y, Seng KW, Demura M (1998) J Mol Struct 446:179
100. Harbison GS, Jelinski LW, Stark RE, Torchia DA, Herzfeld J, Griffin RG (1984) J Magn Reson 60:79
101. Asakawa N, Kuroki S, Ando I, Shoji A, Ozaki T (1992) J Am Chem Soc 114:3261
102. Teng Q, Iqbal M, Cross TA (1992) J Am Chem Soc 114:5312
103. Hong M (2000) J Am Chem Soc 122:3762
104. Kameda T, Takeda N, Kuroki S, Kurosu H, Ando S, Ando I, Shoji A, Ozaki T (1996) J Mol Struct 384:17
105. Tsichiya K, Takashi A, Takeda N, Asakawa S, Kuroki S, Ando I, Shoji A, Ozaki T (1995) J Mol Struct 350:233
106. Kameda T, Ando I (1997) J Mol Struct 412:197
107. Wei Y, Lee D-K, Ramamoorthy A (2001) J Am Chem Soc 123:6118
108. Anzutkin ON, Shekar SC, Levitt MH (1995) J Magn Reson Ser A 115:7
109. Anzutkin ON, Lee YK, Levitt MH (1998) J Magn Reson 135:144

110. Saitô H, Nukada K (1971) J Am Chem Soc 93:1072
111. Saitô H, Tanaka Y, Nukada K (1971) J Am Chem Soc 93:1077
112. Naito A, Tuzi S, Saitô H (1994) Eur J Biochem 224:729
113. Kuroki S, Ando S, Ando I, Shoji A, Ozaki T, Webb GA (1990) J Mol Struct 240:19
114. Fukutani A, Naito A, Tuzi S, Saitô H (2002) J Mol Struct 602/603:491
115. Lumsden MD, Wasylishen RE, Eichele K, Schindler M, Penner GH, Power WP, Curtis RD (1994) J Am Chem Soc 116:1403
116. Hoelger CG, Limbach HH (1994) J Phys Chem 98:11,803
117. Buntkowsky G, Sack I, Limbach HH, Kling B, Fuhrhop J (1997) J Phys Chem B 101:11,265
118. Emshwiller M, Hahn EL, Kaplan DE (1960) Phys Rev Lett 118:414
119. Cleland WW, Kreevoy MM (1994) Science 264:1887
120. Munowitz MG, Griffin RG (1982) J Chem Phys 76:2848
121. Song X, Rienstra CM, McDermott AE (2001) Magn Reson Chem 39:S30
122. Hoelger CG, Aguilar-Parrilla F, Elguero J, Weintraub O, Vega S, Limbach HH (1996) J Magn Reson Ser A 120:46
123. Yamada K, Dong S, Wu G (2000) J Am Chem Soc 122:11,602
124. Dong S, Ida R, Wu G (2000) J Phys Chem A 104:11,194
125. Wu G, Dong S, Ida R, Reen N (2002) J Am Chem Soc 124:1768
126. Heaton NJ, Vold RL, Vold RR (1989) J Am Chem Soc 111:3211
127. Bagno A, Gerard S, Kevalam J, Menna E, Scorrano G (2000) Chem Eur J 6:2915
128. Yamauchi K, Kuroki S, Ando I (2002) J Mol Struct 602:171
129. Kuroki S, Yamauchi K, Ando I, Shoji A, Ozaki T (2001) Curr Org Chem 5:1001
130. Torrent M, Mansour D, Day EP, Morokuma K (2001) J Phys Chem A 105:4546
131. Achlama AM, Kohlschütter U, Haeberlen U (1975) Chem Phys 7:287
132. Lorente P, Shenderovich IG, Golubev NS, Denisov GS, Buntkowsky G, Limbach HH (2001) Magn Reson Chem 39:S18
133. Foces-Foces C, Llamas-Saiz AL, Lorente P, Golubev NS, Limbach HH (1999) Acta Crystallogr Sect C 55:377
134. Gerald R, Bernhard T, Haeberlen U, Rendell J, Opella S (1993) J Am Chem Soc 115:777
135. Reimer JA, Vaughan RW (1980) J Magn Reson 41:483
136. Chesnut DB, Phung CG (1991) Chem Phys Lett 183:505
137. Potrzebowski MJ, Schneider C, Tekely P (1999) Chem Phys Lett 313:569
138. Chu P-J, Potrzebowski MJ, Gao Y, Scott AI (1990) J Am Chem Soc 112:881
139. Gerardy-Montouillout V, Malveau C, Tekely P, Olender Z, Luz Z (1996) J Magn Reson Ser A 123:7
140. Gu ZT, Drueckhammer DG, Kurz L, Liu K, Martin DP, McDermott A (1999) Biochem. 38:8022
141. Usher KC, Remington SJ, Martin DP, Drueckhammer DG (1994) Biochem. 33:7753
142. Arumugam S, Glidewell C, Harris KDM (1992) J Chem Soc Chem Commun 724
143. Olivieri A, Paul IC, Curtin DY (1990) Magn Reson Chem 28:119
144. Medycki W, Reynhardt EC, Latanowicz L (1998) Mol Phys 93:323
145. Llamas-Saiz AL, Foces-Foces C, Sobrados I, Elguero J, Meutermans W (1993) Acta Cryst C49:724
146. Llamas-Saiz AL, Foces-Foces C (1990) J Mol Struct 238:367
147. Le Bas G, Allegret A, Mauguen Y, de Rango C, Bailly M (1980) Acta Crystallogr B36:3007
148. Goward GR, Schnell I, Brown SP, Spiess HW, Kim HD, Ishida, H (2001) Magn Reson Chem 39:S5
149. Leiserowitz L (1976) Acta Cryst B32:775
150. Hayashi S, Umemura J (1974) J Chem Phys 60:2630
151. Gough A, Haq MMI, Smith JAS (1985) Chem Phys Lett 117:389
152. Robyr P, Meier BH, Ernst RR (1991) Chem Phys Lett 187:471
153. Nagaoka S, Terao T, Imashiro F, Sakia A, Hirota N, Hayashi S (1981) Chem Phys Lett 80:580
154. Ernst RR (1990) J Chem Phys 93:1502
155. Heuer A, Haeberlen U (1991) J Chem Phys 95:4201

156. Horsewill AJ, Aibout A (1989) J Phys Condens Matter 1:9609
157. Horsewill AJ, Heidemann A, Hayashi S (1993) Z Phys B90:319
158. Nagaoka S, Terao T, Imashiro F, Saika A, Hirota N, Hayashi S (1983) J Chem Phys 79:4694
159. Nagaoka S, Hirota N, Matsushita T, Nishimoto K (1982) Chem Phys Lett 92:498
160. Meier BH, Graf F, Ernst RR (1982) J Chem Phys 76:767
161. Takwale MG, Pant LM (1971) Acta Cryst B27:1152
162. Furic K (1984) Chem Phys Lett 108:518
163. Meier BH, Meyer R, Ernst RR, Stöckli A, Furrer A, Hälg W, Anderson I (1984) Chem Phys Lett 108:522
164. Meier BH, Meyer R, Ernst RR, Zolliker P, Furrer A, Hälg W (1983) Chem Phys Lett 103:169
165. Stöckli A, Furrer A, Schoenenberger C, Meier BH, Ernst RR, Anderson I (1986) Physica B 136:161
166. Nagaoka S, Terao T, Imashiro F, Hirota N, Hayashi S (1984) Chem Phys Lett 108:524
167. Grabowski SJ, Krygowski TM (1988) Chem Phys Lett 151:425
168. Scheubel W, Zimmerman H, Haeberlen U (1988) J Magn Reson 80:401
169. Haeberlen U (1992) Mol Phys 76:157
170. Idziak S, Pislewski N (1987) Chem Phys 111:439
171. Szverenyi NM, Bax A, Maciel GE (1983) J Am Chem Soc 105:2579
172. Titman JJ, Luz Z, Spiess HW (1992) J Am Chem Soc 114:3756
173. Larsen RG, Lee YK, Yang JO, Luz Z, Zimmermann H, Pines A (1995) J Chem Phys 103:9844
174. Olender Z, Reichert D, Muller A, Zimmermann H, Poupko R, Luz R (1996) J Magn Reson 120:31
175. Detken A, Zimmermann H, Haeberlen U, Luz Z (1997) J Magn Reson 126:95
176. Bernhard T, Zimmermann H, Haeberlen U (1990) J Chem Phys 92:2178
177. Aliev AE, MacLean EJ, Harris KDM, Kariuki BM, Glidewell C (1998) J Phys Chem B 102:2165
178. Serrano-González H, Harris KDM, Wilson CC, Aliev AE, Kitchin SJ, Kariuki BM, Bach-Vergés M, Glidewell C, MacLean EJ, Kagunya WW (1999) J Phys Chem B 103:6215
179. Ferguson G, Gallagher JF, Glidewell C, Low JN, Scrimgeour SN (1992) Acta Cryst C48:1272
180. Aliev AE, Atkinson CE, Harris KDM (2002) J Phys Chem B 106:9013
181. Latanowicz L, Reynhardt EC (1994) Ber Bunsenges Phys Chem 98:818
182. Long JR, Sun BQ, Bowen A, Griffin RG (1994) J Am Chem Soc 116:11,950
183. Kitchin SJ, Ahn S, Harris KDM (2002) J Phys Chem A 106:7228
184. Gu ZT, Ebisawa K, McDermott A (1996) Solid State Nucl Magn Reson 7:161
185. Smith AE (1952) Acta Crystallogr 5:224
186. Harris KDM, Thomas JM (1990) J Chem Soc Faraday Trans 86:2985
187. Heaton NJ, Vold RL, Vold RR (1989) J Magn Reson 84:333
188. Aliev AE, Smart SP, Harris KDM (1994) J Mater Chem 4:35
189. Emsley JW, Smith JAS (1961) Trans Faraday Soc 57:1233
190. Chiba T (1965) Bull Chem Soc Japan 38:259
191. Zussman A (1973) J Chem Phys 58:1514
192. Mantsch HH, Saito H, Smith ICP (1977) Prog NMR Spectr 11:211
193. Das TP (1957) J Chem Phys 27:763
194. Das TP (1961) J Chem Phys 35:1897
195. Truter MR (1967) Acta Crystallogr 22:556
196. Elcombe MM, Taylor JC (1968) Acta Crystallogr A 24:410
197. Kristensen JH, Hoatson GL, Vold RL (1998) Solid State Nucl Magn Reson 13:1
198. Baldy A, Elguero J, Faure R, Pierrot M, Vincent E-J (1985) J Am Chem Soc 107:5290
199. Elguero J, Fruchiel A, Pellegrin V (1981) J Chem Soc Chem Commun 1207
200. Faure R, Vincent E-J, Elguero J (1983) Heterocycles 20:1713
201. Aguilar-Parrilla F, Scherer G, Limbach HH, Foces-Foces C, Cano FH, Smith JAS, Toiron C, Elguero J (1994) J Am Chem Soc 114:9657
202. Meschede L, Limbach HH (1991) J Phys Chem 95:10,267
203. Kawada A, McGhie AR, Labes MM (1970) J Chem Phys 52:3121

204. Mirza P, Agarmal P, Gupta RC (1969) Indian J Pure Appl Phys 12:716
205. Martinez-Carrera S (1966) Acta Crystallogr 20:783
206. Craven BM, McMullan RK, Bell JD, Freeman HC (1977) Acta Crystallogr B33:2585
207. Ueda T, Masui H, Nakamura N (2001) Solid State Nucl Magn Reson 20:145
208. Song X, McDermott AE (2001) Magn Reson Chem 39:S37
209. Garcia MA, Lopez C, Peters O, Claramunt RM, Klein O, Schagen D, Limbach HH, Foces-Foces C, Elguero J (2000) Magn Reson Chem 38:604

Structure and Bonding, Vol. 108 (2004): 55–96
DOI 10.1007/b14137

Crystal Engineering Using Multiple Hydrogen Bonds

Andrew D. Burrows

Department of Chemistry, University of Bath, Claverton Down, Bath BA2 7AY, United Kingdom
E-mail: a.d.burrows@bath.ac.uk

Abstract Crystal engineering is the branch of supramolecular chemistry concerned with the design and synthesis of extended structures with predictable form and function. In this chapter, the use of hydrogen bonds to generate one-, two- and three-dimensional structures is discussed, with the different strategies employed compared. The review concentrates on systems in which two or more hydrogen bonds link components together, and extended structures based on both one and two components are highlighted. Parallels are drawn between crystal engineering using purely organic components, and the more recent extension to the inclusion of coordination and organometallic complexes.

Keywords Crystal engineering · Extended structures · Networks · Hydrogen bonding · Self-assembly

1 Introduction 56

2 The Language of Crystal Engineering 57

2.1 Supramolecular Synthons 57
2.2 Hydrogen Bond Donors and Acceptors 57
2.3 Tectons and Janus Molecules 58
2.4 Graph Set Notation 58
2.5 Secondary Interactions 58
2.6 Tapes, Chains, Ribbons, Sheets and Networks 59
2.7 Bifunctional Ligands 59
2.8 Complementarity and One and Two Component Systems 59

3 Systems Based on DA-AD Interactions 60

3.1 DA-AD Synthons Involving Pairs of OH···O Hydrogen Bonds 60
3.2 DA-AD Synthons Involving Pairs of NH···O Hydrogen Bonds 66
3.3 DA-AD Synthons Involving Pairs of OH···N Hydrogen Bonds 70
3.4 DA-AD Synthons Involving Pairs of NH···N Hydrogen Bonds 70
3.5 Non-Self-Complementary DA-AD Interactions 75
3.6 DA-AD Synthons Involving Weaker Hydrogen Bonds 75

4 Systems Based on DD-AA Interactions 76

4.1 One Component Systems 76
4.2 Guanidinium Nitrate and Guanidinium Sulfonates 77

4.3 Bis(Amidinium) Dicarboxylates 80
4.4 Thiourea Dicarboxylate Complexes and Related Systems 81

5 **Systems Based on ADA-DAD Interactions** 83

6 **Systems Based on DDA-AAD, DDA′-A′DD and DDD-AAA Interactions** 87

7 **Systems in Which Molecules Are Linked by Four or More Hydrogen Bonds** 89

8 **Outlook** 90

9 **References** 90

1 Introduction

The emergence of supramolecular chemistry [1] is arguably the most important development in chemistry over the past few decades. A significant feature of this approach has been a change in emphasis from the discrete molecule to the 'supermolecule', and this has led to what philosophers would term a *gestalt shift* in the way in which solid state structures are visualised. No longer is attention limited to molecular structure, but the manner in which the molecules pack is also of interest, and means for trying to control this assembly are of increasing importance. Crystal engineering [2, 3] is the name given for this attempted control of solid state structure. The term 'crystal engineering' was first used by Schmidt more than 30 years ago in discussion of photodimerisation reactions in crystalline cinnamic acids [4], though it is only in the past 10–15 years that interest in the area has blossomed and the underlying concepts developed. Modern crystal engineering is an interdisciplinary subject, with input from and implications for organic, inorganic, organometallic, theoretical and materials chemistry, in addition to biology, crystallography and crystal growth. A useful modern definition is that provided by Desiraju [5] who defined crystal engineering as '*the understanding of intermolecular interactions in the context of crystal packing and the utilisation of such understanding in the design of new solids with desired physical and chemical properties*'. Since many of the bulk properties of molecular materials are dictated by the manner in which the molecules are ordered in the solid state, it is clear that an ability to control this ordering would afford control over these properties. Although the prediction and design of solid state structures is fundamental to crystal engineering, it is not synonymous with crystal structure prediction, which is a far more precise and difficult to achieve challenge [6]. However, since extended structures may have repeat distances commensurate with the unit cell lengths of the crystal, and their group symmetries are a subgroup of the space group of the final crystal, they can be important models for crystal structure prediction. One problem for the design of extended structures is the potential existence of polymorphs which may involve different packing motifs, though use of strong intermolecular interactions would be expected to reduce the frequency of this. The existence of polymorphs within a given system can be assessed through a comparison of X-ray powder diffraction data with those simulated from single crystal analyses.

Crystal engineering relies on non-covalent forces to achieve the organisation of molecules and/or ions in the solid state. Much of the initial work on purely organic systems focussed on the use of hydrogen bonds [7, 8], though the more recent extension to inorganic systems has seen the coordination bond also emerge as a powerful crystal engineering tool [9–11]. The object of this review is to survey recent progress in crystal engineering using multiple hydrogen bonds. Hydrogen bonds [12, 13] are employed in crystal engineering studies for three main reasons: they are relatively strong, directional and able to act in concert with each other. Most crystal engineering studies have employed what can be termed 'traditional' hydrogen bonds – X-H···Y interactions in which both X and Y are electronegative atoms, normally N or O. Weaker hydrogen bonds such as C-H···O [14] and X-H···Cl-M [15] interactions have also been used in this context, though a discussion of these is beyond the scope of this review. The review will concentrate on the progress made in preparing one-, two- and three-dimensional hydrogen-bonded networks, and will seek to draw parallels between studies carried out using purely organic systems, and those including coordination and organometallic complexes. It is therefore arranged on the basis of the type of hydrogen bonding motif employed. First of all, some of the terminology used in crystal engineering studies is discussed.

2 The Language of Crystal Engineering

2.1 Supramolecular Synthons

By analogy with the synthons of organic synthesis, Desiraju [16] introduced the term '*supramolecular synthon*' to describe the '*structural units within supermolecules which can be formed and/or assembled by known or conceivable synthetic operations involving intermolecular interactions*'. Such supramolecular synthons (often abbreviated to synthons in the literature, and hereafter) can involve two identical or different components (Fig. 1).

2.2 Hydrogen Bond Donors and Acceptors

In the majority of examples in this review the hydrogen bond donors are XH bonds and the acceptors are the lone pairs on electronegative atoms, though other sources of electron density such as π-bonds can also act as hydrogen bond acceptors. The interactions between molecules or ions can be described in terms of the number and orientation of hydrogen bond donors (D) and acceptors (A) present. For example the carboxylic acid dimer in Fig. 1a can be denoted DA-AD (or DA:AD, or DA=AD) whereas the fragment in Fig. 1c can be denoted ADA-DAD (or ADA≡DAD).

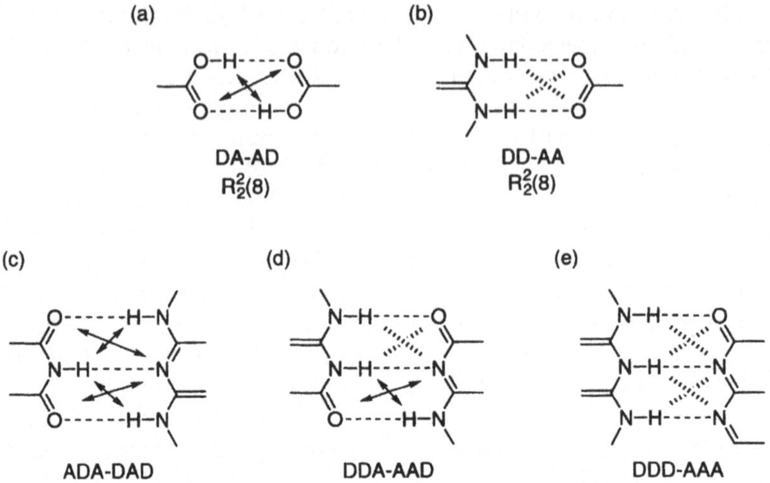

Fig. 1 Examples of supramolecular synthons based on two and three hydrogen bonds, with favourable secondary interactions shown as *dashed lines*, and unfavourable secondary interactions shown as *double headed arrows*

2.3
Tectons and Janus Molecules

The word *tecton* (from the Greek *tekton*, builder) was coined by Wuest and co-workers [17] to describe a molecule whose interactions are dominated by particular associative forces that induce the self-assembly of an organised network with specific structural features. Thus in the following discussion, the individual molecular or ionic components that together generate a hydrogen-bonded network can be described as tectons. The term *Janus molecule* (from the Roman god Janus, who possesses faces in the front and back of his head) was first used by Lehn and co-workers [18] to describe a molecule containing two hydrogen bonding faces.

2.4
Graph Set Notation

This methodology, developed by Etter and co-workers, enables hydrogen bonding motifs within three-dimensional structures to be described simply, thus facilitating comparison between structures. Each motif is described as a ring (R), a chain (C), a non-cyclic dimer (D) or intramolecular (S). The degree (in parentheses) is the number of atoms in the repeat unit, and the super- and sub-scripts represent the number of hydrogen bond acceptors and donors in the motif respectively [19,20].

2.5
Secondary Interactions

Hydrogen bonds are predominantly electrostatic in nature, so when multiple hydrogen bonds are present secondary interactions need to be considered in addition to the primary attractive interactions [21,22]. These secondary interactions can be either attractive or repulsive, as illustrated for the systems containing two and three hydrogen bonds illustrated in Fig. 1. The energy of a secondary interaction has been calculated as approximately 7 kJ mol^{-1} for a system containing three parallel hydrogen bonds [23]. As a result of these secondary interactions, the DD-AA motif is expected to be more favourable than the DA-AD motif, whereas for systems containing three hydrogen bonds, DDD-AAA is more favourable than DDA-AAD, which in turn is more favourable than ADA-DAD.

2.6 Tapes, Chains, Ribbons, Sheets and Networks

The terms tape, chain and ribbon are often used interchangedly for one-dimensional structures, though several authors have attempted to define each more precisely [7, 24]. In this review, a chain is defined as an infinite one-dimensional structure in which components are linked by single hydrogen bonds whereas a tape is defined as an infinite one-dimensional structure in which components are linked by more than one hydrogen bond, with the hydrogen bonds approximately co-planar. Sheets are infinite two-dimensional structures, whereas networks are infinite three-dimensional structures. The terms α-, β- and γ-network are also used in the literature to describe one-, two- and three-dimensional structures respectively. The precise definition of these relates to the degrees of translational symmetry present, with α-networks having one, β-networks having two, and γ-networks having three [25].

2.7 Bifunctional Ligands

Ligands that have been designed to have two separate functions can be described as bifunctional [23, 26]. In the context of crystal engineering, bifunctional ligands are those capable of binding a metal atom while retaining a face capable of hydrogen bonding. Incorporation of transition metal ions in predetermined positions may allow for the tailoring of materials with specific magnetic, optical or electronic properties. In addition, metal ions can provide templates with different geometries to those available in purely organic systems, for example octahedral and square-planar, that enable different supramolecular structures to be constructed.

2.8 Complementarity and One and Two Component Systems

Crystal engineering strategies employing hydrogen bonds generally involve either one or two component systems. In one component systems, the tectons are self-complementary and hence possess the potential to self-assemble into the desired one-, two- or three-dimensional structure. Examples include the DA-AD interaction present in carboxylic acid dimers and the AADD-DDAA interaction present in ureidopyrimidones [27], though Janus molecules containing two complementary faces are also possible. One-component systems have the advantage of simplicity, though may suffer from unfavourable secondary interactions and Coulombic repulsions in ionic systems. In two-component systems the tectons are complementary to each other, and the interactions between tectons can be enhanced by attractive secondary interactions and/or Coulombic attractions, such as in the DD-AA interactions of guanidinium sulfonates [28]. In order for two-component systems to be used they must be able to co-crystallise, either from the solution, which implies the solubilities of the components must be similar, or by grinding the components together [19].

3 Systems Based on DA-AD Interactions

3.1 DA-AD Synthons Involving Pairs of OH···O Hydrogen Bonds

The archetypal self-complementary supramolecular synthons are carboxylic acid dimers, which form $R_2^2(8)$ rings (Fig. 1a). Extension from the discrete zero-dimensional dimer into one-, two- and three-dimensional structures can be facilitated by the incorporation of more than one carboxylic acid group into a molecule. Hence simple dicarboxylic acids such as terephthalic acid and isophthalic acid typically exhibit tape structures whereas tricarboxylic acids such as trimesic acid form sheet structures (Fig. 2).

The crystal structure of trimesic acid was reported in 1969 [29], and contains the anticipated hexagonal 'chicken wire' motif, though the presence of large voids is countered by interpenetration. Non-interpenetrating trimesic acid structures have been observed through the inclusion of guest molecules such as isooctane [30] or large aromatics such pyrene [31, 32], though these structures can include other hydrogen bonding motifs in addition to $R_2^2(8)$ rings, such as $R_4^4(12)$ 'expanded dimers' in which water or alcohol molecules mediate between the carboxylic acid groups (Fig. 3). Indeed, this motif has been shown to be robust, and observed in the crystal structures of a range of hydrated dicarboxylic acids [33]. The crystal structure of tributyltrimesic acid forms a non-interpenetrating 'chicken wire' sheet structure with the butyl groups filling the voids. However other trialkyltrimesic acids do not form analogous structures, and the $R_2^2(8)$ rings are either observed in combination with other motifs, or not observed at all [32].

For tetracarboxylic acids, the relative orientation of the hydrogen bonding groups is the dominant factor in determining the supramolecular structure adopted. In methanetetracarboxylic acid **1** [34] and adamantane-1,3,5,7-tetracarboxylic acid **2** [35], the carboxylic acid groups are arranged tetrahedrally, and the resultant networks are diamondoid (Fig. 4), with threefold and fivefold interpenetration respectively, though this can be reduced by substitution [36]. In contrast, the structures of 1,1′-biphenyl-2,2′,6,6′-tetracarboxylic acid **3** [37] and the porphyrin-based tetracarboxylic acid **4** [38] contain square-based grids whereas that of bis(4-butyl-3,5-dicarboxyphenyl)acetylene **5** contains a similar network to that of trimesic acid, with a third of the $R_2^2(8)$ rings replaced by covalent acetylene linkers [32].

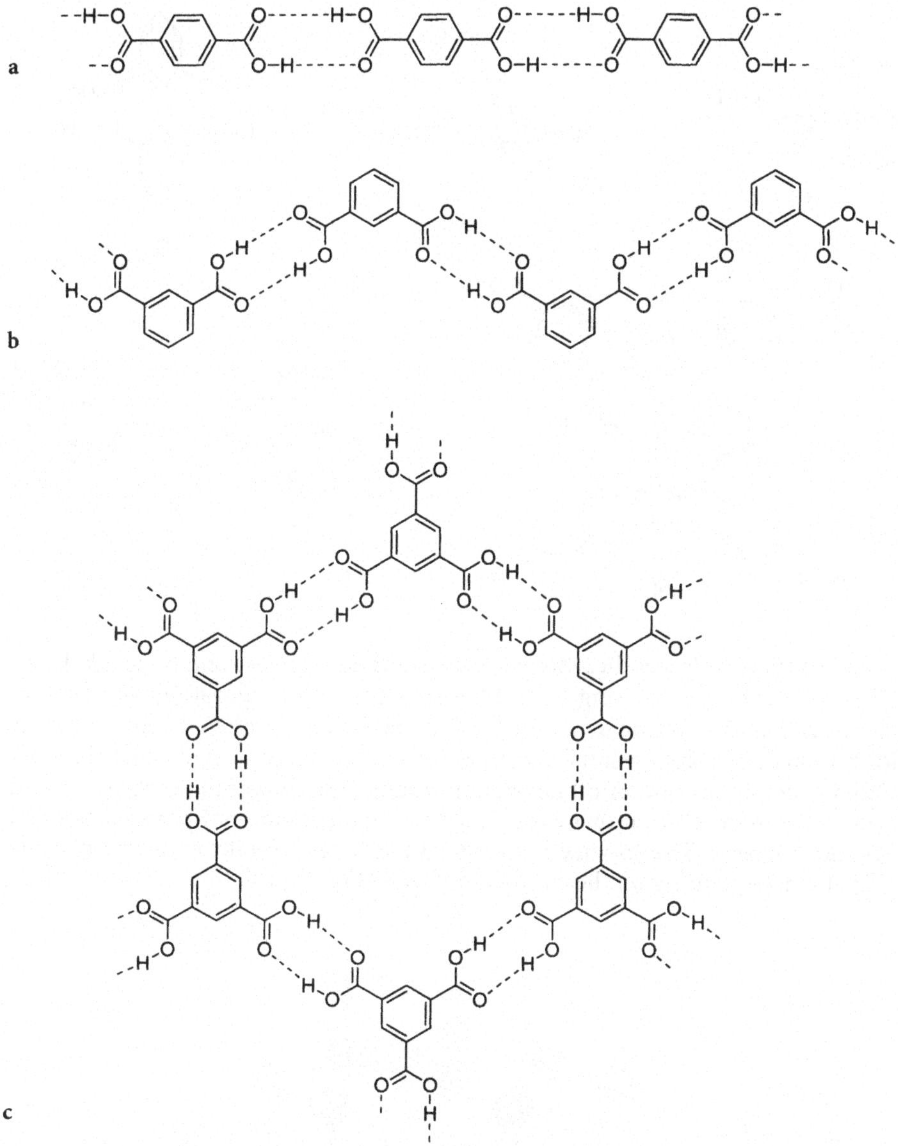

Fig. 2 **a** The linear tape structure of terephthalic acid. **b** The zigzag tape structure of isophthalic acid. **c** The 'chicken wire' sheet structure of trimesic acid [29]

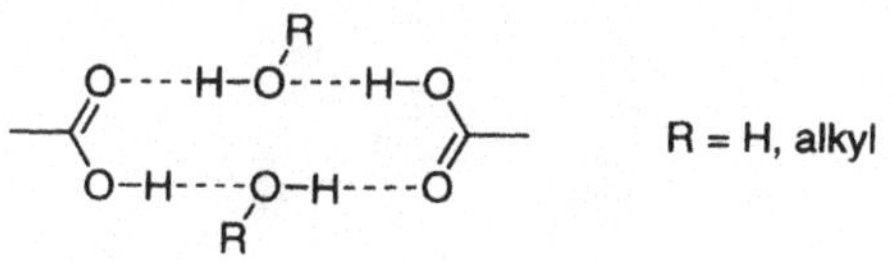

Fig. 3 The $R_4^4(12)$ 'expanded carboxylic acid dimer' motif

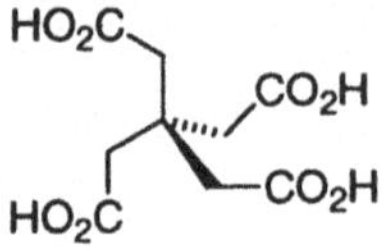

1

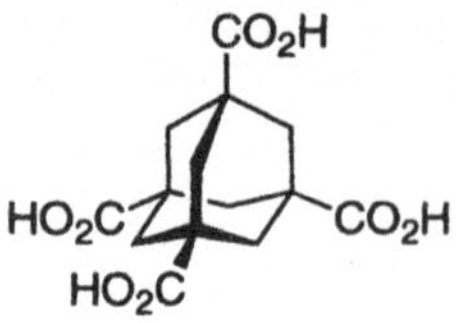

2

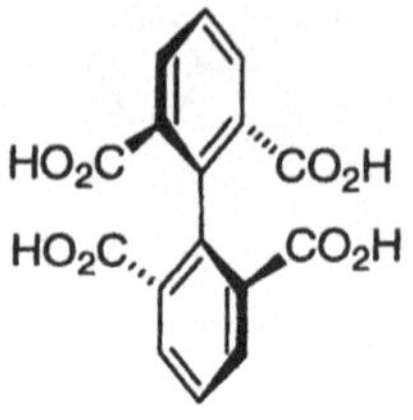

3

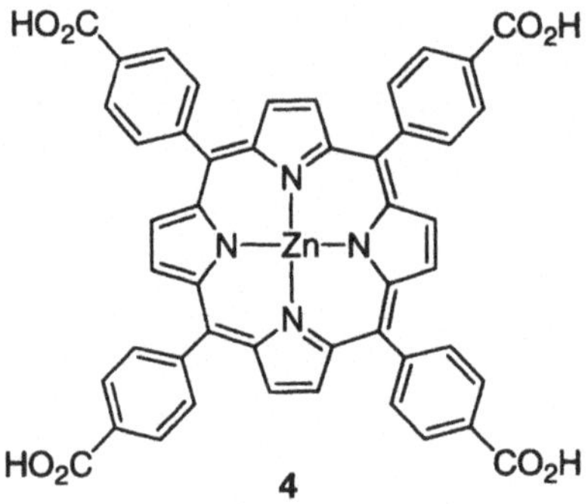

4

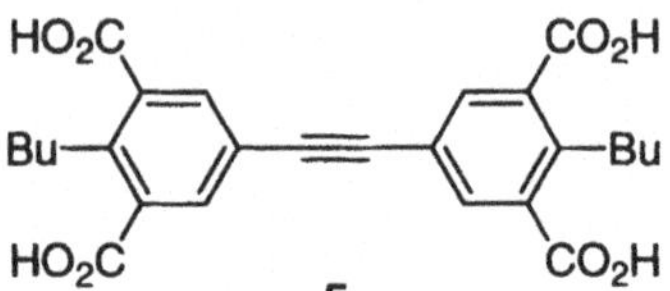

5

N-Phenylpyrrole-2,5-dicarboxylic acid **6** and derivatives containing alkyl substituents in the phenyl ring form tape structures. Incorporation of a further carboxylic acid functionality into the 4-position of the phenyl group induces formation of two-dimensional sheets, which contain large solvent-filled channels [39]. Co-crystallisation of dibenzylammonium salts containing carboxylic acid groups together with a crown ether leads to the formation of hydrogen-bonded pseudorotaxanes. The crown ether rings can be located on the main chain or the side chain, depending on the cation employed (Fig. 5) [40].

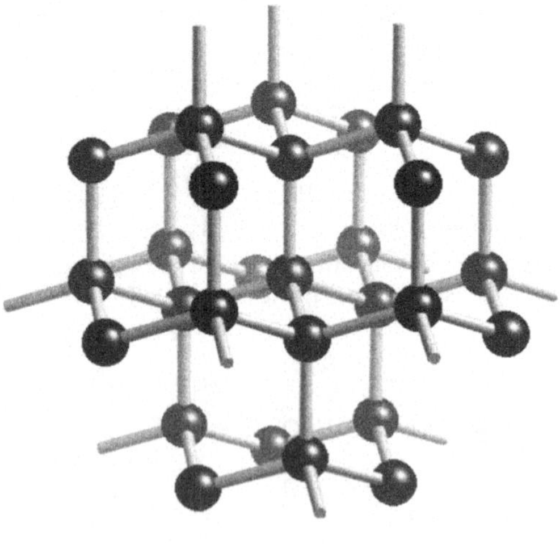

Fig. 4 The three-dimensional diamondoid array

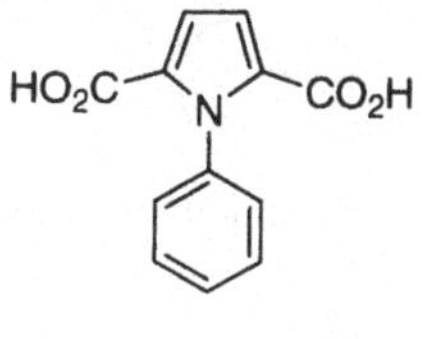

6

A metal centre can be incorporated into structures based on the carboxylic acid dimer synthon by inclusion of an additional functionality capable of acting as a ligand. Thus tapes can be formed in the solid state by using a ligand containing two carboxylic acid groups, such as fumaric acid in the structure of $[Fe(CO)_4\{\eta^2\text{-}C_2H_2(CO_2H)_2\text{-}trans\}]$ [41], or two ligands each bearing one carboxylic acid group, such as $PPh_2CH_2CO_2H$ in the structure of $[PdCl_2(PPh_2CH_2CO_2H)_2]$ [42]. Pyridine ligands containing one or two carboxylic acids have also attracted recent attention. The palladium nicotinic and isonicotinic acid complexes *trans*-$[PdCl_2(n\text{-}NC_5H_4CO_2H)_2]$ (n=3 or 4) form the anticipated one-dimensional tapes [43], though cyclic arrays have been observed in platinum-PPh_3 complexes [44]. The isonicotinic acid dimer can be considered as a hydrogen-bonded analogue of 4,4′-bipyridyl (Fig. 6), and may therefore be expected to have a similar rich

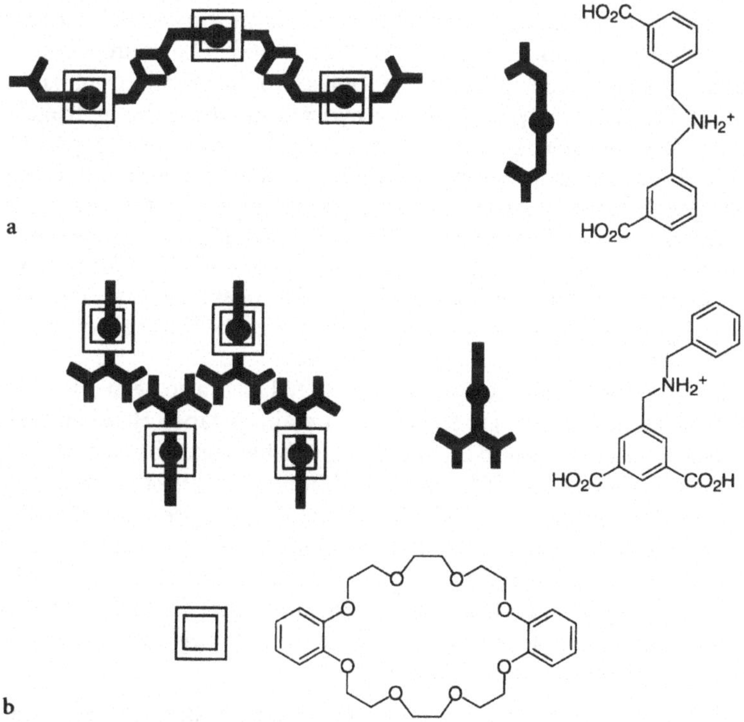

Fig. 5a, b Schematic structures of hydrogen-bonded pseudorotaxanes, with crown ether rings located: **a** on the main chain; **b** on the side chains [40]

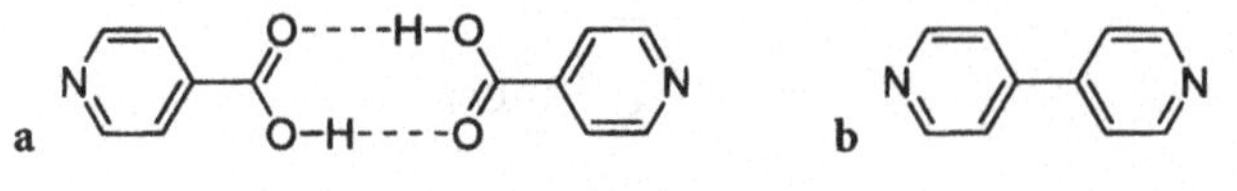

Fig. 6 a The isonicotinic acid dimer, depicting its structural similarity with b 4,4′-bipyridine

structural chemistry to this ligand. The structure of $[Ni(\mu\text{-}SCN)_2(4\text{-}NC_5H_4CO_2H)_2]$ suggests this will be the case, with one-dimensional $\{Ni(\mu\text{-}SCN)_2\}_\infty$ coordination polymers linked through carboxylic acid dimers to give a sheet structure which allows the incorporation of guest molecules such as biphenyl into the cavities. These cavities can be lengthened by using a longer dimer such as that derived from 3-(4-pyridinyl)-2-propenoic acid or an isonicotinic acid-fumaric acid 2:1 trimer [45]. One-dimensional copper(I) halide chains have also been linked into sheets by carboxylic acid dimers, in this case involving coordinated 6-methylnicotinic acid molecules [46].

Braga, Grepioni and co-workers have examined organometallic sandwich compounds in which carboxylic acids have been included into cyclopentadienyl or benzene ligands. There would appear to be a delicate balance between formation of dimers linked by two DA-AD interactions and of hydrogen-bonded tapes. Dimers have been observed for $[Fe(\eta^5\text{-}C_5H_4CO_2H)_2]$ [47] and $[Cr(\eta^6\text{-}C_6H_5CO_2H)_2]$ [48], whereas tapes have been observed for $[Co(\eta^5\text{-}C_5H_4CO_2H)_2]PF_6$ [49], $[Cr(\eta^6\text{-}C_6H_5CO_2H)_2]PF_6$ [48] and $[Fe\{\eta^5\text{-}C_5H_3(1\text{-}Me)(3\text{-}CO_2H)\}_2]$ [50]. In the latter case, tape formation is believed to result from minimisation of steric repulsions, through adoption of a staggered geometry. The counter-ion can also play a structural role, where present – in contrast to the hexafluorophosphate salt, $[Co(\eta^5\text{-}C_5H_4CO_2H)_2]Cl$ crystallises as a hydrate in which the hydrogen bonds between carboxylic acid groups are replaced by ones involving the anions and water, whereas in the presence of included urea the predominant motif is DA-AD interactions between carboxylic acid groups and urea molecules [49]. The chromium trimesic acid complex $[Cr\{\eta^6\text{-}C_6H_3(CO_2H)_3\}(CO)_3]$ crystallises with an equivalent of di(*n*-butyl)ether, which blocks one of the hydrogen bonding faces on the complex, preventing formation of a sheet structure. The other carboxylic acid groups interact to give tapes reminiscent of those observed in isophthalic acid [51].

The alkyl ligands $CH_2C_6H_4CO_2H$ and CH_2CO_2H have also been introduced into metal complexes and can lead to structures containing tapes. However these tapes are not always observed due to competition with hydrogen bonding to anions or included solvent molecules [52]. Indeed, although simple supramolecular synthons can lead to predictable solid state structures, the presence of such groups is not sufficient for such structures to form. A study of the Cambridge Structural Database has showed that only in approximately a third of the cases where the carboxylic acid dimers can theoretically form are they actually observed [53]. There are a number of reasons why tape formation through the carboxylic dimer motif may not be observed, including:

- Formation of competing intramolecular hydrogen bonds. Etter established empirically that strong hydrogen bond donors and acceptors preferentially form intramolecular hydrogen bonds if the system allows such a possibility

[19]. Intramolecular OH···O hydrogen bonds are observed in the structure of both maleic acid and its iron complex [Fe(CO)$_4${η^2-$C_2H_2(CO_2H)_2$-*cis*}] [41].
- Formation of hydrogen bonds to included molecules or anions. In the presence of additional hydrogen bond donors or acceptors, other motifs may be more favourable, such as that in which a pyridine nitrogen atom acts as acceptor (see below).
- Space-filling problems. Generally crystal structures are thermodynamically favoured if the packing efficiency is high [54]. In very regular structures unfavourable voids can be avoided by interpenetration of open networks. In less regular structures this may not be possible without introducing unfavourable steric interactions, hence either efficient space-filling or hydrogen bonding potential has to be compromised, though a vector-based approach has shown how close-packing and hydrogen bonding can work together to give efficient packing [55].
- Formation of discrete dimers as opposed to infinite tapes as, for example, in [Fe(η^5-$C_5H_4CO_2H)_2$] [47].
- Modification of the hydrogen bonding groups. A carboxylic acid can be readily deprotonated to form a carboxylate, which contains a hydrogen bonding face that is no longer self-complementary. The crystal structure of a cobalt complex based on dicarboxylic acid-functionalised 2,2′-bipyridine ligands was observed to contain both carboxylic acid and carboxylate functionalities, which hydrogen bond together to generate interpenetrating networks [56].
- Formation of alternative hydrogen bonding patterns. Although less common than the $R_2^2(8)$ motif, carboxylic acids can adopt the C(4) motif. This has been shown to be important for sterically hindered 2,6-disubstituted benzoic acids [57].

Although most studies involving pairs of OH···O hydrogen bonds have used carboxylic acids, other functional groups with this potential have also been investigated. Hydrogen bonding between hydrogen phosphate anions of the general formula $RPO_2(OH)^-$ generally involve strong symmetrical single hydrogen bonds and/or DA-AD interactions. A study of nitrilotri(methylphosphonic acid) revealed monoanions adopt self-complementary three-dimensional hexagonal architectures whereas dianions give tapes [58]. DA-AD interactions have also been observed in the structures of complexes of a bis(phosphonomethyl)azacrown ligand, and again these serve to link the complexes into tapes in the solid state (Fig. 7) [59].

In metal-containing systems, synthons involving coordinated water or ammine ligands are possible. For example, hydrogen bonding between coordinated water and pyridine oxide ligands in a $R_2^2(8)$ motif has been observed to link cobalt 4,4′-bipyridine dioxide coordination polymers into sheets [60].

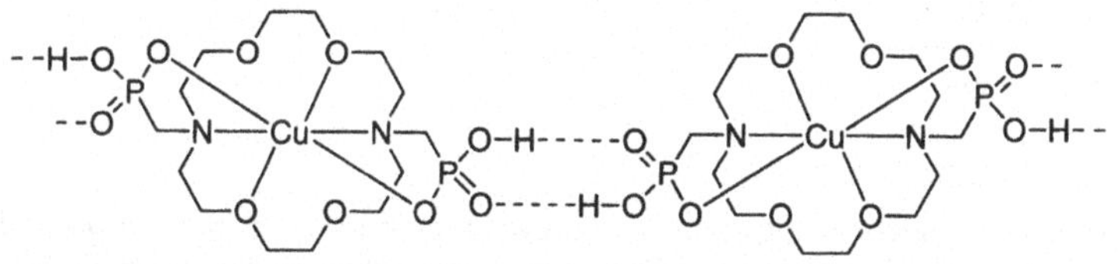

Fig. 7 Tape formation in a copper bis(phosphonomethyl)azacrown complex [59]

3.2
DA-AD Synthons Involving Pairs of NH···O Hydrogen Bonds

Amides contain the same potential hydrogen bonding face as carboxylic acids, while eliminating the problem of deprotonation. The crystal structures of amides have been studied in detail [24, 61, 62], and for primary amides the DA-AD synthon, formed between faces comprising the carbonyl and the *syn* NH group, has been demonstrated to be the most common pattern. The *anti* NH group typically forms C(4) motifs which in combination with the $R_2^2(8)$ ring can lead to tapes (Fig. 8). These primary amide tapes are not disrupted by the presence of the hydrogen bond donors and acceptors present in sulfamide groups. Indeed, each sulfamide group forms hydrogen bonds to four adjacent molecules giving a sheet structure, which is linked into the third dimension by the amide···amide interactions [63]. A range of cubanecarboxamides show the dimers linked in a less favourable manner due to a steric mismatch between the cubyl group (5.4 Å) and the translation tape repeat length (5.1 Å) [64].

Whitesides and co-workers have studied the structures of symmetrically substituted diketopiperazines (or piperazinediones) such as 7 which contain two DA faces [65, 66]. These molecules form hydrogen-bonded tapes in the solid state, with the planarity of the tapes dependent on the conformation adopted by the diketopiperazine ring. The rigidity of the molecules, together with observed hydrogen bonding, make this system particularly appealing for packing calculations, and a simulated annealing Monte Carlo-based procedure was used to correctly predict the crystalline structures of several examples. Chiral centres have been included into diketopiperazines, and though the fine details of packing can vary, tape formation is maintained in *meso*-, *rac*- and (*S*,*S*) isomers [67, 68]. Diketopiperazine tapes can be linked into corrugated sheets by incorporation of additional hydrogen bonding groups. For example, in the crystal structure of the cyclic dipeptide of (*S*)-aspartic acid **8** the tapes are connected by carboxylic acid dimers [69]. Another means of linking the tapes into sheets is co-crystallisation with a dicarboxylic acid, such as in the 1:1 adduct between the cyclic dipeptide of glycine and 2,5-dibromoterephthalic acid [70].

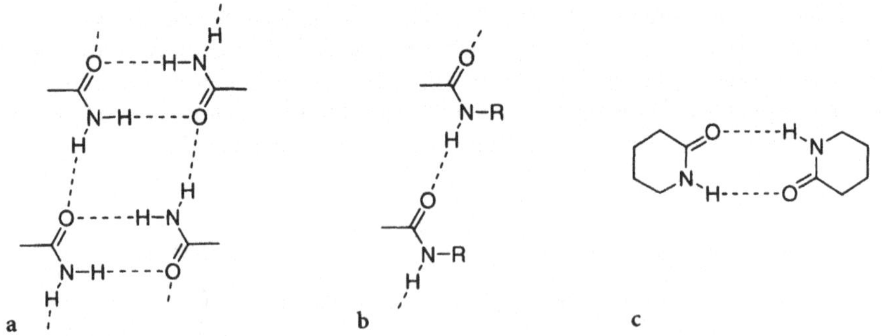

Fig. 8 Common hydrogen bonding patterns observed for amides: **a** the tape structure formed by primary amides involving $R_2^2(8)$ and C(4) motifs; **b** chains formed by non-cyclic secondary amides; **c** the dimer formed by cyclic amides [62]

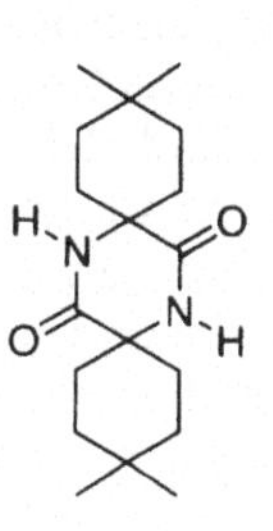

7

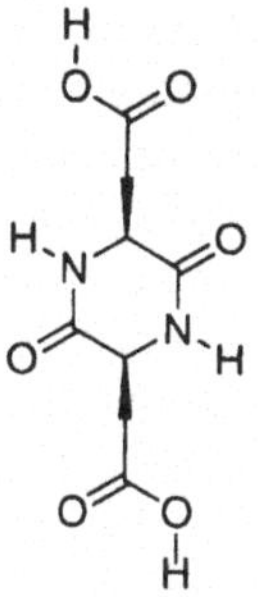

8

Cyclic ureas and thioureas also crystallise to give tapes, though as expected the NH···O hydrogen bonds in ureas lead to a more robust motif than the NH···S hydrogen bonds in thioureas [71]. Dialkyl glycolurils form similar tape structures (Fig. 9), with substantial twisting around the bridgehead observed. This twisting makes the molecules chiral, and this chirality is transmitted to the tapes [72, 73].

Hydrogen bond interactions between the $NH_2C(O)CO_2^-$ anions in oxamate salts have been shown to be dependent on the nature of the cations, with $R_2^2(8)$ and $R_2^2(10)$ motifs between like faces and $R_2^2(9)$ motifs between unlike faces all having been observed. In the two homodimeric systems, single hydrogen bonds link the dimers into tapes or sheets, whereas for the heterodimeric system, the $R_2^2(9)$ motifs lead directly to tapes [74].

The tendency of 2-pyridones to form hydrogen-bonded dimers (Fig. 8c) was exploited by Wuest and co-workers in the formation of diamondoid structures based on rigid tetrapyridones such as **9** [17]. Crystallisation of **9** from butyric acid/methanol/hexane gave the anticipated diamondoid network, which is sevenfold interpenetrating [75] and contains butyric acid molecules present in the pores.

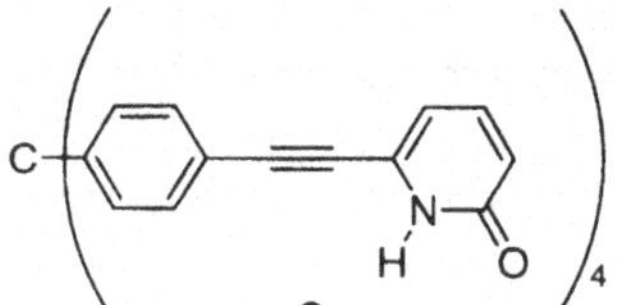

9

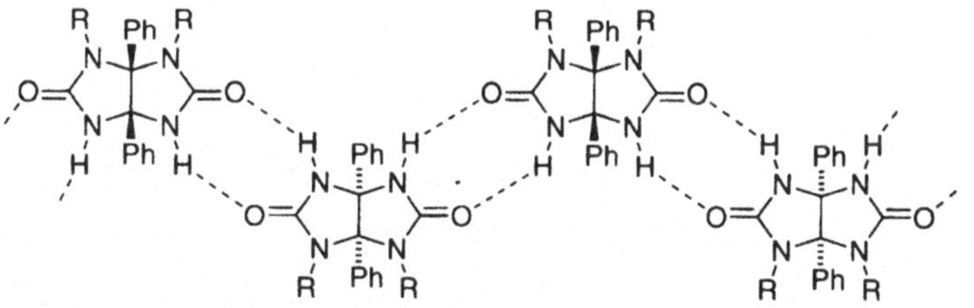

Fig. 9 Chiral tapes formed by substituted glycolurils [72]

2-Pyridone co-crystallises in a 2:1 ratio with dicarboxylic acids to give structures in which two 2-pyridone molecules interact to give hydrogen-bonded dimers. These dimers are linked by OH···O and CH···O hydrogen bonds, in which the dicarboxylic acid acts as donor to give infinite tapes [76,77]. None of these structures include the carboxylic acid dimer $R_2^2(8)$ motif, which is consistent with reports that in co-crystals between acids and amides the acid OH group tends to hydrogen bond to the amide oxygen atom. Similar tape structures are observed when the 2-pyridone dimer is replaced by a phenazine molecule [78], however co-crystals with a 1:1 stoichiometry consist of 2-pyridone dimers linked together by dicarboxylic acid dimers [79].

Pioneering work by Etter and co-workers on co-crystals involving a range of molecules such as acyclic imides and ureas were fundamental in deriving empirical rules on hydrogen bonding in organic solid state structures [19,80]. Symmetrically disubstituted ureas generally form hydrogen-bonded tapes involving $R_2^1(6)$ motifs in which both NH groups on one molecule interact with the carbonyl group on a second, giving a repeat unit of 4.60 Å [81], and these tapes can be linked into sheets by co-crystallisation with a α,ω-dinitrile [82]. Symmetrically disubstituted oxamides also form hydrogen-bonded tapes, in this instance involving $R_2^2(8)$ motifs and a repeat unit of 5.05 Å [83]. Addition of carboxylic acid and pyridine groups to ureas and oxamides gives complementary molecules such as **10** and **11** which can be co-crystallised, with hydrogen bonds between the carboxylic acid and pyridine moieties linking the urea and/or oxamide tapes into sheets. When two complementary urea compounds or two complementary oxamide compounds were co-crystallised the anticipated sheet structures were formed. Co-crystallisation of the urea derivative **10** with the oxamide derivative **11** also gave a sheet structure, but with a repeat distance intermediate (4.87–4.88 Å) between those observed for ureas and oxamides [25].

Molecules containing ADA faces, designed for use in two-component systems (see below), also contain by necessity self-complementary DA faces, and though not all hydrogen bond acceptors will be satisfied, these molecules can self-assemble in the absence of complementary molecules that carry DAD faces. Thus cyanuric acid forms tapes through complementary DA-AD interactions (Fig. 10), which are linked into sheets either via single NH···O hydrogen bonds, or by pairs of hydrogen bonds involving a solvent such as DMF [84]. Cyanurate complexes such as *trans*-[Cu(cyanurate-*N*)$_2$(NH$_3$)$_2$] form sheet structures in which cyanurate

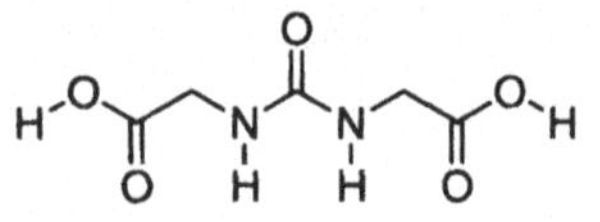

10

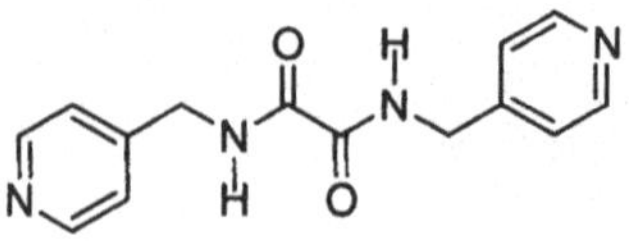

11

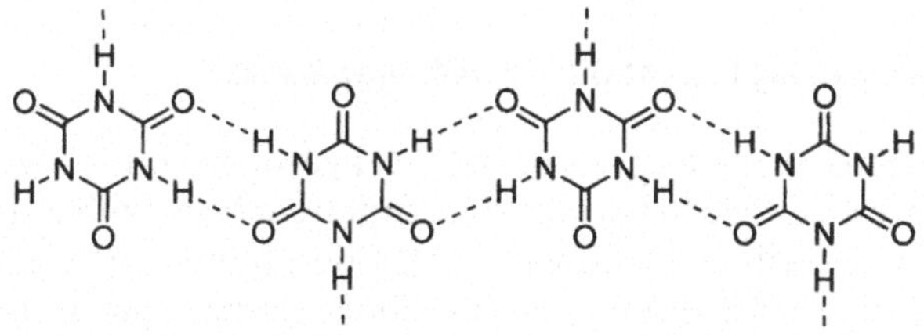

Fig. 10 Hydrogen-bonded tapes formed by cyanuric acid in the absence of a compound bearing the complementary DAD face [84]

tapes are linked together through the copper centres [85]. Interactions between thiobarbiturate ligands have also been employed to give tapes, and in gold(I) complexes these have been used in combination with Au···Au interactions [86].

Incorporation of metal centres into amide hydrogen-bonded arrays has also been achieved using bifunctional ligands such as nicotinamide. The silver complexes $[Ag\{NC_5H_3(6\text{-Me})(3\text{-CONH}_2)\}_2]X$ ($X=NO_3$, OTf) form structures in which amide···amide $R_2^2(8)$ dimers link the cations into pairs, which are linked into tapes through single NH···O hydrogen bonds and interactions involving the anions. In contrast the $R_2^2(8)$ dimers in $[Ag\{NC_5H_4(4\text{-CONH}_2)\}_2]OTf$ link the cations into tapes that are connected into sheets through hydrogen bonds to the anions and included water molecules [87], and weaker attractive forces such as Ag···Ag interactions have been used to connect tapes into sheets [88]. Coordination of mutually *cis* nicotinamide or isonicotinamide ligands to palladium or platinum can lead to one-dimensional zigzag tapes as in $[Pt(PEt_3)_2\{NC_5H_4(4\text{-CONH}_2)\}_2](NO_3)_2$ [89] and $[Pd(dppp)\{NC_5H_4(3\text{-CONH}_2)\}_2](OTf)_2$ [90], though dimers linked by NH···O hydrogen bonds into ladders have also been observed [90]. The rhodium(III) complex $[Rh(\eta^5\text{-}C_5Me_5)\{NC_5H_4(4\text{-CONH}_2)\}_3](OTf)_2$ adopts infinite interwoven strands based on amide···amide dimers [89]. Use of *N*-methylnicotinamide leads to hydrogen bonding patterns based on the C(4) motif as opposed to the $R_2^2(8)$ motif since adoption of the *anti* conformation removes the potential for DA-AD formation.

A more complex ligand is the diurea-substituted phenanthroline ligand **12**. The complex $[Fe(\mathbf{12})_3]Br_2$ adopts a layer structure in the solid state containing both enantiomers and channels of approximately 10 Å diameter. Individual cations are connected through $R_2^2(8)$ motifs in addition to single NH···O hydrogen bonds [91]. Other metal complexes linked by DA-AD interactions include copper complexes of 2-hydroxyquinoxaline **13**, in which the hydrogen bonding combines with π···π stacking [92] and ferrocene-pyrrole hybrids such as **14** [93].

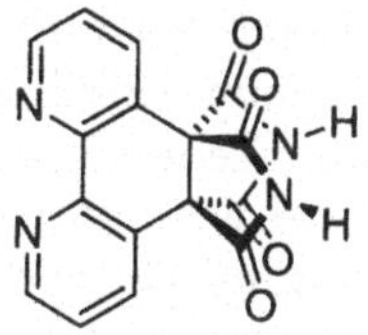

12

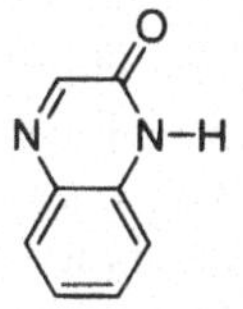

13

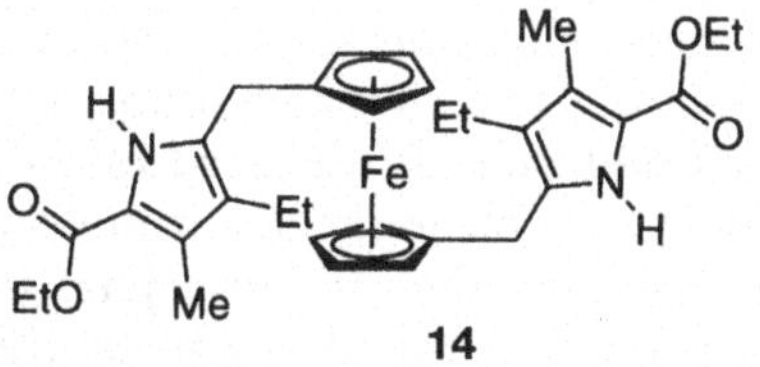

14

3.3
DA-AD Synthons Involving Pairs of OH···N Hydrogen Bonds

DA-AD interactions involving pairs of OH···N hydrogen bonds have received less study, though are observed between oxime functionalities with a similar relative frequency to carboxylic acid dimers [53]. Compounds such as 1,1′-bicyclohexylidene-4,4′-dione dioxime form one-dimensional tapes in the solid state through formation of oxime···oxime $R_2^2(6)$ motifs (Fig. 11) [94, 95]. Similar tapes are observed between the cations in linear two-coordinate silver(I) complexes of pyridyl-functionalised oximes [96], and the motif is not disrupted even in the presence of additional hydrogen bond acceptors on the ligands [97]. However, free pyridyl oximes tend to form infinite chains linked by single OH···N hydrogen bonds as opposed to discrete dimers [98, 99].

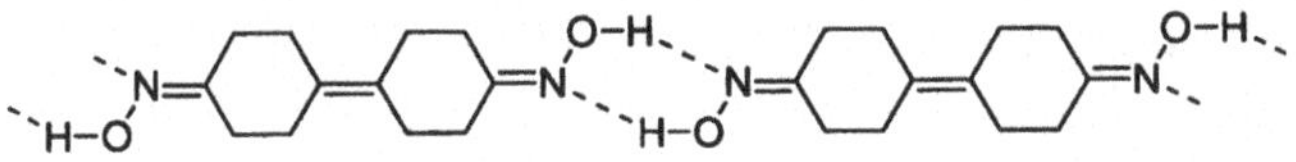

Fig. 11 Hydrogen-bonded tapes formed by 1,1′-bicyclohexylidene-4,4′-dione dioxime, based on $R_2^2(6)$ motifs [94]

3.4
DA-AD Synthons Involving Pairs of NH···N Hydrogen Bonds

2-Aminopyridines can dimerise via $R_2^2(8)$ motifs, and this interaction can be used as the basis for forming extended structures through use of molecules containing two such groups. *N,N′*-Bis(2-pyridyl)aryldiamines from hydrogen-bonded tapes when the arylaminopyridines adopt *E* conformations, but singly-hydrogen-bonded chains when they adopt *Z* conformations [100]. An alternative approach involves the use of 2-aminopyrimidine **15** which has two available DA faces. Although 2-aminopyrimidine might be anticipated to give a tape structure in the solid state, its crystal structure reveals the presence of two-dimensional 'slabs' due to one of the DA faces acting as hydrogen bond donor and acceptor to two different molecules [101]. The combination of two 2-aminopyrimidine moieties into the same molecule has afforded both tapes and sheets, with a cofacial arrangement of the 2-aminopyrimidine moieties a necessary condition for sheet formation and the use of all hydrogen bonding groups in the array, as observed for **16** [102]. Adjusting the angle between the pyrimidine moieties by employing a linking adamantylidene group led to the formation of pleated sheets, though the size of the alkylidene substituent is crucial, as use of *tert*-butylcyclohexylidene led to some of the hydrogen bonds becoming geometrically unfeasible [103]. In contrast, the chiral bis-2-aminopyrimidine **17** led to a structure containing helical columns [104]. The anthracene-substituted molecule **18** was used in order to generate structures that were anticipated to be able to include π-molecular guests via π···π interactions. The crystal structure of a 1:1 co-crystal between **18** and phenazine supported this assertion, showing a sheet structure held together by DA-AD interactions, with

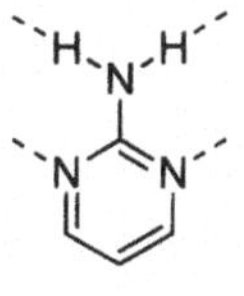

15

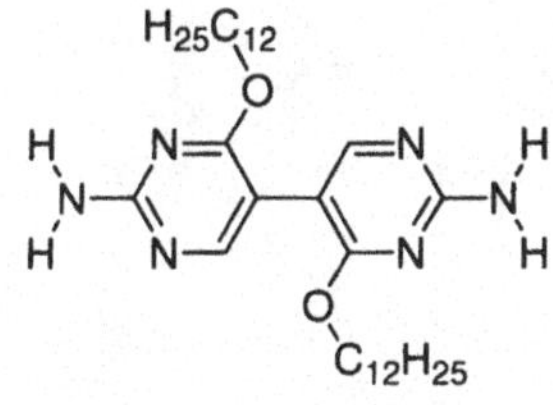

16

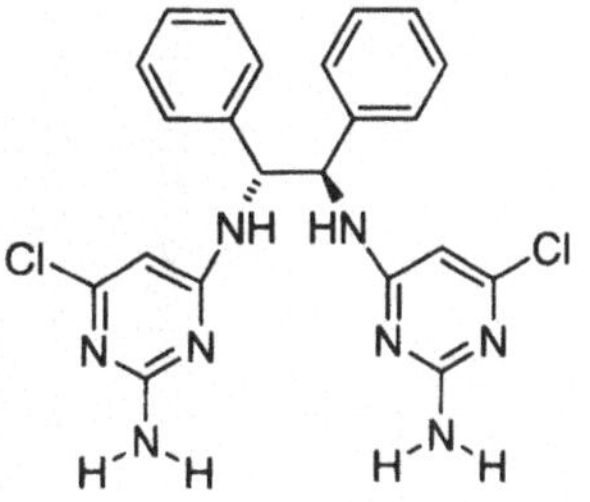

17

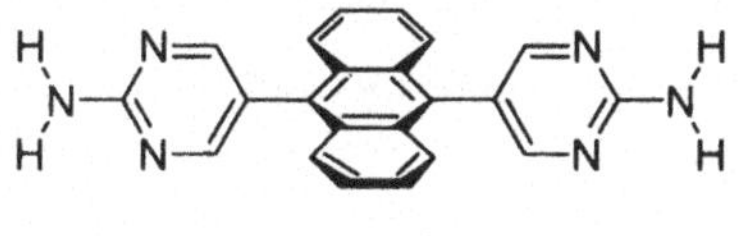

18

phenazines incorporated into the cavities between pairs of parallel anthracene units [105].

Inclusion of 2-aminopyrimidine-type moieties into terpyridines to give ligands such as **19** leads to the possibility of forming transition metal complexes bearing four DA faces. The crystal structure of $[Co(\mathbf{19})_2](PF_6)_2$ revealed a two-dimensional grid in the solid state, with the cations linked by the anticipated $R^2_2(8)$ motifs, though in the tetrafluoroborate analogue, the hydrogen bonds are only fully satisfied in one direction. This difference is believed to relate to the size of the anions, with the hexafluorophosphate ions being more complementary to the size of the cavities in the grid [106]. Extension to a more complex ligand such as **20** affords the possibility of generating two-dimensional arrays of [2×2] grid-type tetranuclear complexes. The crystal structure of $[Co_4(\mathbf{20})_4](BF_4)_8$ showed DA-AD hydrogen bonding to be only present in one dimension, affording tapes of grids (Fig. 12). Adoption of this structure rather than the two-dimensional array avoids the formation of large cavities [107].

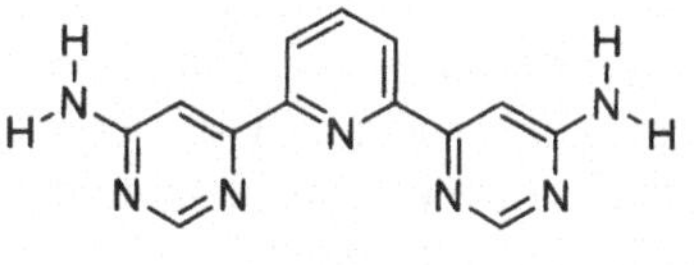

19

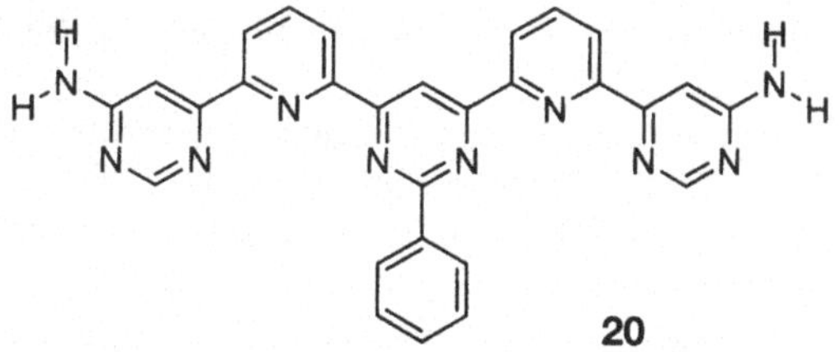

20

2,2′-Biimidazole (H_2bim) forms linear tapes in the solid state through formation of $R^2_2(10)$ motifs [108]. Inclusion of bulky substituents in the 4- and 5-positions to give **21** prevents the hydrogen-bonded tapes from lying flat, and helical columns were observed instead [109]. The compound H_2bim in various degrees of deprotonation has also been used to generate different types of network when coordinated to a metal centre [110]. When monodeprotonated it becomes self-

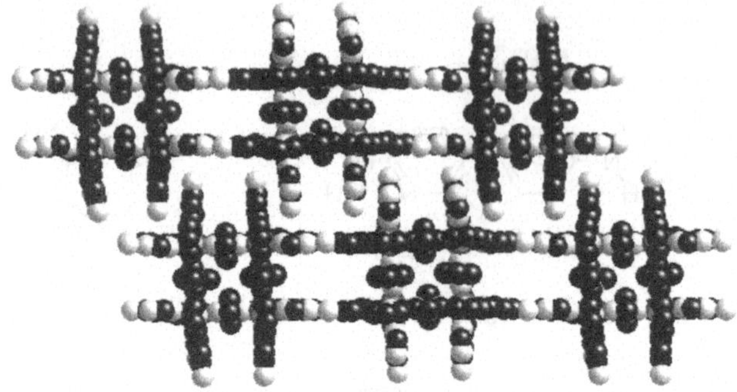

Fig. 12 Packing of $[Co_4(20)_4]^{8+}$ cations into tapes [107]

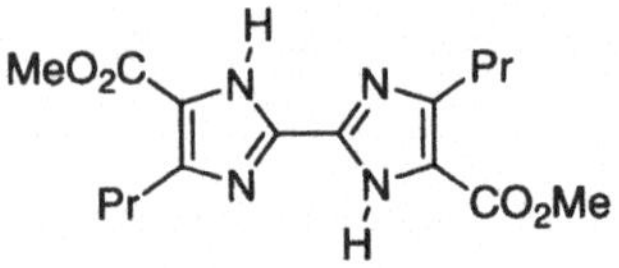

21

complementary, hence 2,2′-biimidazolate complexes have the potential to self-assemble into extended structures. The complexes *trans*-$[Ni(Hbim)_2(^tBupy)_2]$ and $[Ni(Hbim)_2(bipy)]$ both form one-dimensional tapes, which are linear and zigzag respectively (Fig. 13) [111]. The self-assembly of anions in compounds incorporating $[Ni(Hbim)_3]$ depends on the cation, with two-dimensional 'honeycomb' sheets (Fig. 13) in which all the DA faces are involved in hydrogen bonding possible, though not always observed due to solvent inclusion [112, 113]. Such honeycomb sheets are observed with the neutral racemic compound $[Ru(Hbim)_3]$, though the structure adopted by $[Co(Hbim)_3]$ is dependent on both the solvent and the chirality of the metal centre, with crystals grown from methanol/ethyl acetate giving honeycomb sheets, those from DMF/water giving (10,3) nets, and crystals of either the Δ or Λ isomer giving helical one-dimensional tapes, with one DA face on each molecule blocked by interactions with ethanol or 2-propanol [114] [111, 115].

In a similar manner to the self-assembly observed with molecules containing ADA faces, molecules containing DAD faces also contain by necessity self-complementary DA faces, and can self-assemble in the absence of complementary molecules that carry ADA faces. Thus, for example, the rigid tetrahedral molecule **22** crystallises to give a porous three-dimensional network in which each molecule is hydrogen-bonded to eight others. This network is sufficiently robust for the dioxane solvate to retain crystallinity on removal of most of the guests [116]. Porphyrin-based compounds such as **23** form square-based hydrogen-bonded arrays. The 22 Å-wide cavities are filled in part by interpenetration, though the structure occupies only 65% of the crystal volume [117]. 2,4-Diamino-6-(4-pyridyl)-1,3,5-triazine **24** forms a sheet structure in the solid state, with half of the molecules employing three DA faces and the other half two DA faces in the

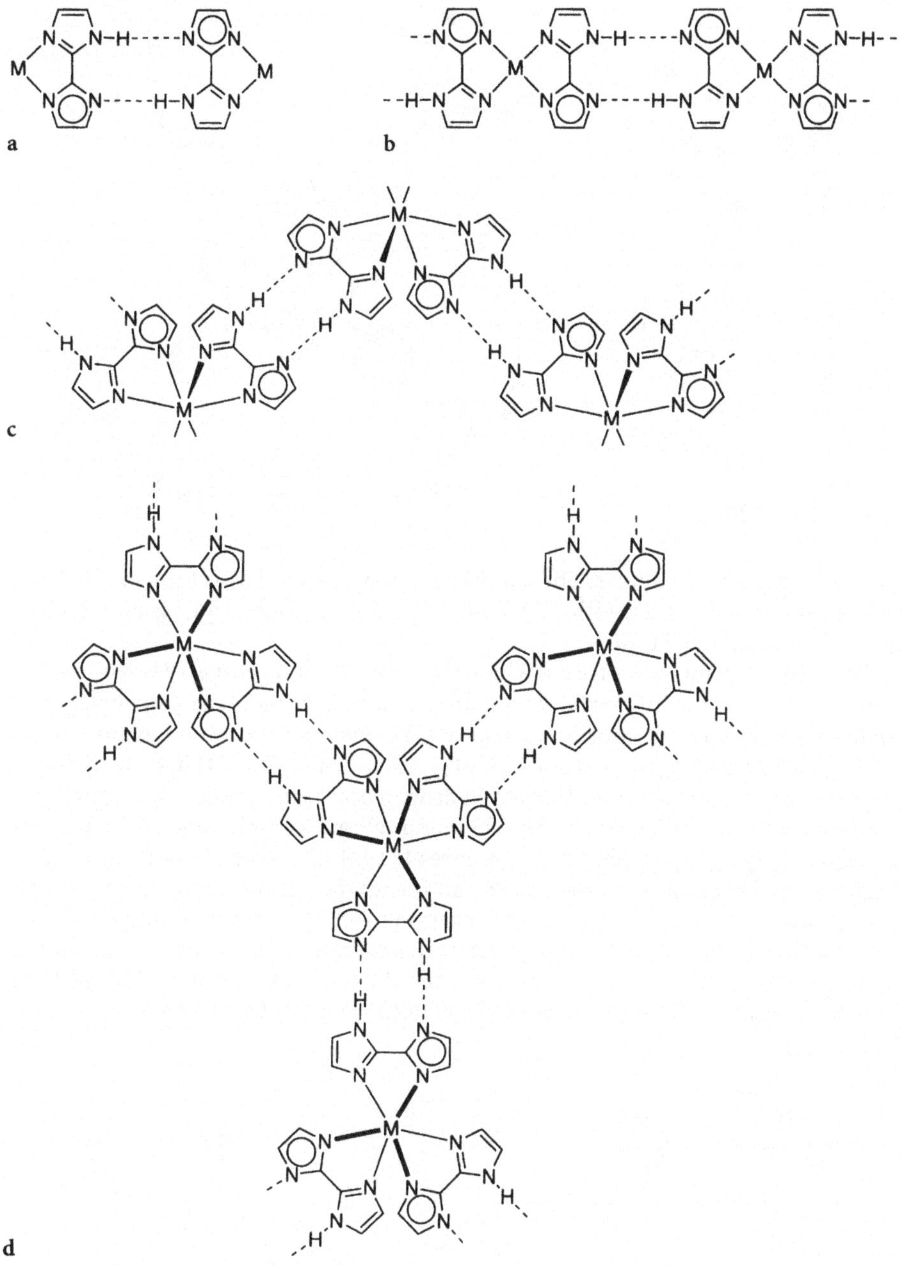

Fig. 13a–d Solid state structural types formed by biimidazolate complexes: **a** dimers; **b** linear tapes; **c** zigzag tapes; **d** hexagonal sheets [111]

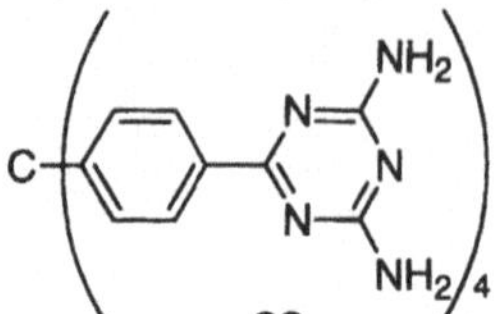

22

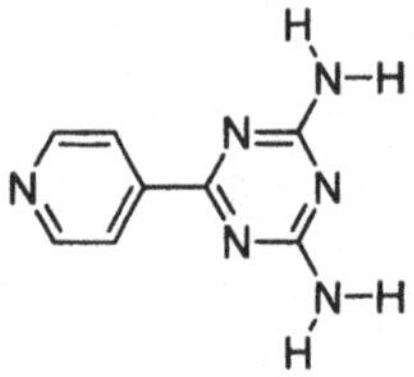

24

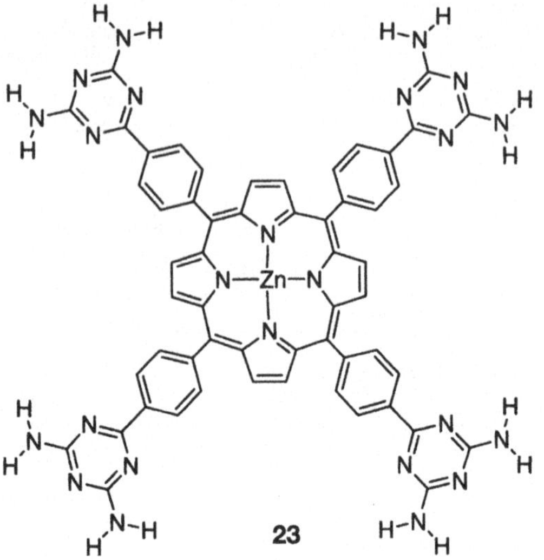

23

hydrogen-bonded structure. The copper complex *cis*-[Cu(**24**)$_2$(*O*-dmso)$_4$](ClO_4)$_2$ forms corrugated sheets, with each copper(II) centre connecting tapes of hydrogen-bonded ligands [118].

For transition metal centres in which the coordination geometry is readily distorted, the existence of particular hydrogen bonding patterns can dictate the metal geometry. Three-coordinate copper(I) centres are observed in complexes such as [{Cu(cnge)}$_2$(μ-pyridazine)$_2$](BF_4)$_2$ **25**, in which DA-AD hydrogen bonding between ligand faces and DD-AA hydrogen bonding between ligand and anion faces enforce a sheet structure, whereas analogues employing nitrile ligands instead of cnge give tetrahedral complexes [119, 120]. Neutral bis(thiosemicarbazidato)nickel complexes such as **26** can form tapes in the solid state based on DA-AD interactions, though their observation in analogues containing alkyl substituents in the endocyclic amine group is dependent on the orientation adopted by this group. In these cases, rotation around the C-N bond removes the DA face, thus rendering $R^2_2(8)$ motifs impossible, with C(4) chains being observed instead [121].

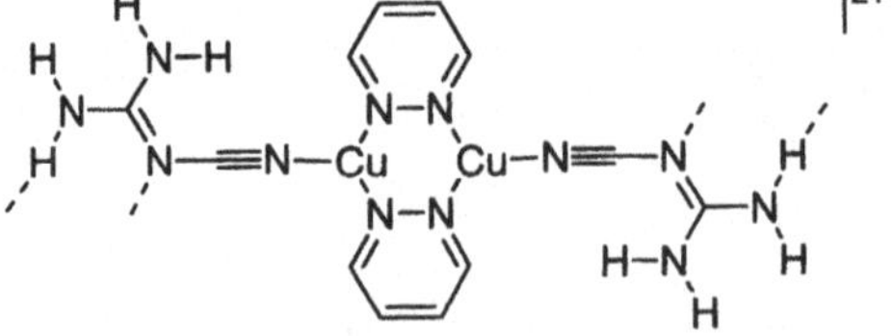

25

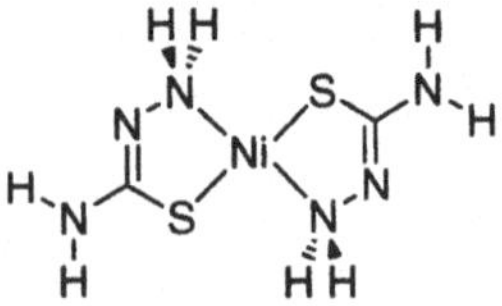

26

3.5
Non-Self-Complementary DA-AD Interactions

Although both carboxylic acids and aminopyridines are self-complementary, these groups preferentially hydrogen bond to each other, giving $R_2^2(8)$ rings (Fig. 14). This has been rationalised on the basis of the best hydrogen bond donor (the hydroxyl) interacting preferentially with the best hydrogen bond acceptor (the pyridine nitrogen atom), the principle of which is formalised in Etter's rules [19]. The use of bis(aminopyridines) together with dicarboxylic acids enables formation of tapes in the solid state, though if there is a good correspondence between the spacer lengths in the two molecules discrete 1:1 adducts can result instead [122, 123]. Co-crystallisation of 2-aminopyrimidine with 1,4-naphthalenedicarboxylic acid was observed to give a sheet structure containing cross-linked tapes based on $R_2^2(8)$ rings [124].

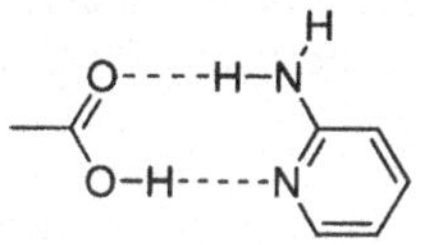

Fig. 14 Robust hydrogen bonding observed between a carboxylic acid and an aminopyridine

3.6
DA-AD Synthons Involving Weaker Hydrogen Bonds

While the focus of this review is on combinations of 'traditional' hydrogen bonds, weaker interactions can also be used instead of, or together with, these interactions. For example, tapes have been observed in the crystal structures of derivatives of benzimidazolene-2-thione **27** [125], though the weakness of the NH···S

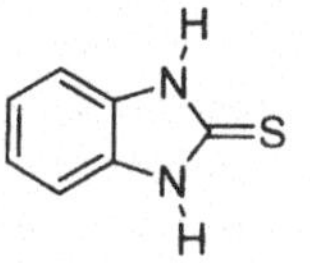

27

hydrogen bonding may contribute to the relatively high observation of polymorphism in this system. In addition, combinations of NH···N and CH···O hydrogen bonds in a $R_2^2(7)$ motif have been shown to be robust enough to generate helices in 5,5-diethylbarbituric acid-hexamethylenetetramine co-crystals [126], whereas the hydrogen bonding in pyrazinecarboxylic acids also gives $R_2^2(7)$ motifs, in this case with a OH···N hydrogen bond being supported by a CH···O hydrogen bond, in preference to carboxylic acid dimers (Fig. 15) [127].

Alcohols and amines are not normally regarded as good hydrogen bond acceptors, however interactions between vicinal diols and vicinal diamines lead to well-defined assemblies in which diol and diamine molecules are linked by

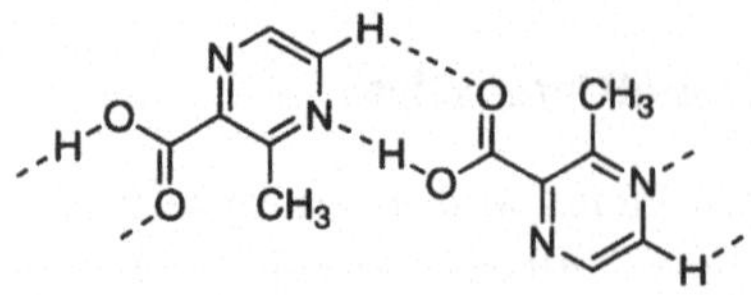

Fig. 15 Hydrogen bonding observed in pyrazinecarboxylic acids, involving an $R_2^2(7)$ motif [127]

two hydrogen bonds, typically but not exclusively of the form DA-AD. These units are further connected into helicate columns via additional hydrogen bonds [128, 129].

Bifurcated hydrogen bonds between dipyridinium cations and coordinated dithiooxalates have been used to assemble tapes [130] with similar geometry to those in which coordinated halides act as acceptors [15]. Replacement of dipyridinium with 4-carboxypiperidinium gave a structure in which the anions are linked by carboxylic acid dimer assembled dications. Similar bifurcated hydrogen bonds were observed in the structures of bipyridinium salts of anilic acids [131].

4 Systems Based on DD-AA Interactions

4.1 One Component Systems

One means of attempting to increase the crystal engineering success rate is to make the supramolecular synthon used more robust. This can be achieved by designing the synthon so that the secondary interactions are attractive. Since the DD-AA interaction (Fig. 1b) involves greater stabilisation than the DA-AD interaction, use of this motif would be expected to lead to greater frequency of occurrence. Anions containing both carboxylate and urea groups have been prepared and shown to crystallise as tapes [132]. However, these interactions are not observed in the presence of a competitive solvent. More robust aggregation is observed in the structures of guanidinium-carboxylate zwitterionic derivatives such as the guanidinonicotinate **28**, which forms tape structures from water. The intramolecular hydrogen bond ensures the molecule adopts a planar configuration [133]. Tapes were also observed in the structure of 3-amidinium benzoate **29** [134], and in this case they are cross-linked by additional NH···O hydrogen bonds to give sheets.

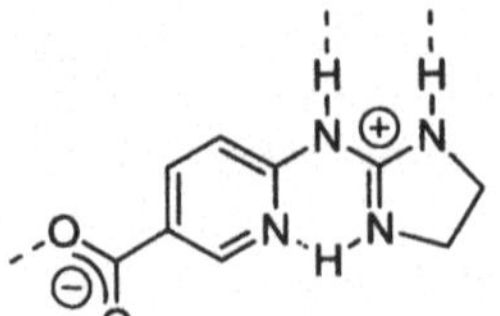

28

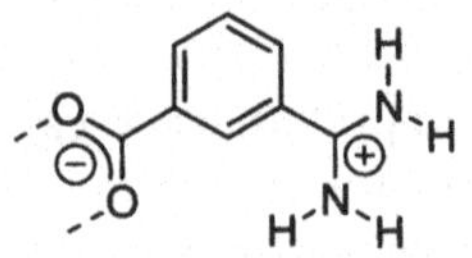

29

4.2 Guanidinium Nitrate and Guanidinium Sulfonates

The archetypal two-component system involving DD-AA interactions is that of guanidinium nitrate [135]. Both cation and anion have three equivalent hydrogen bonding faces and these interact to give hexagonal sheets with the ions linked through $R_2^2(8)$ rings. The use of a sulfonate instead of nitrate allows a variety of alkyl and aryl substituents to be introduced into the arrays (Fig. 16a) and the effects of these have been studied extensively by Ward and co-workers [136, 137]. They have found that with relatively small substituents (<~4.4 Å), the guanidinium sulfonates form bilayer structures in which the substituents on each sheet are orientated to the same side (Fig. 17a). However, an increase of substituent size disfavours the interdigitation of these layers, and consequently larger substituents give rise to continuous single layer stacking in which the substituents are orientated to both sides of a given sheet (Fig. 17b). The guanidinium sulfonate (GS) hydrogen-bonded network is tolerant to a wide range of sulfonate substituents due

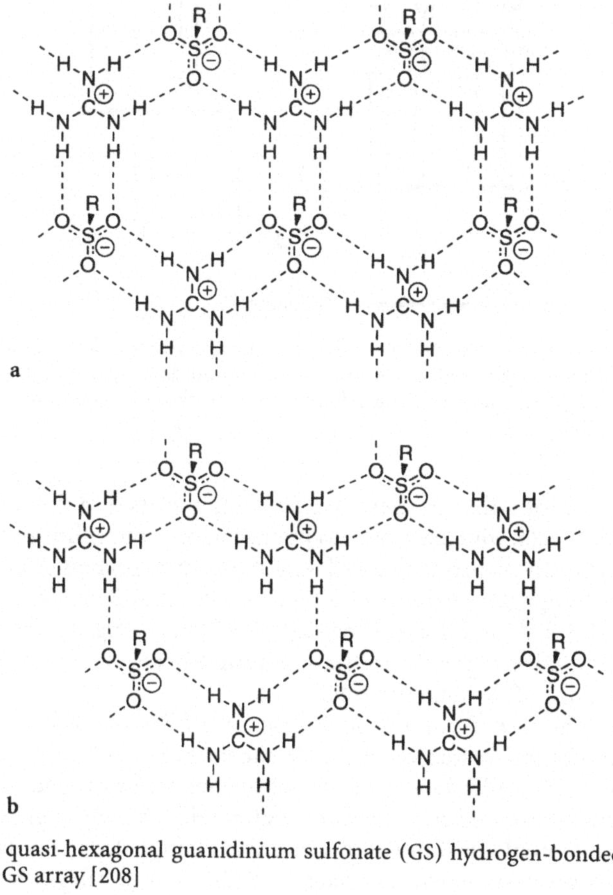

Fig. 16 **a** The quasi-hexagonal guanidinium sulfonate (GS) hydrogen-bonded array. **b** The shifted ribbon GS array [208]

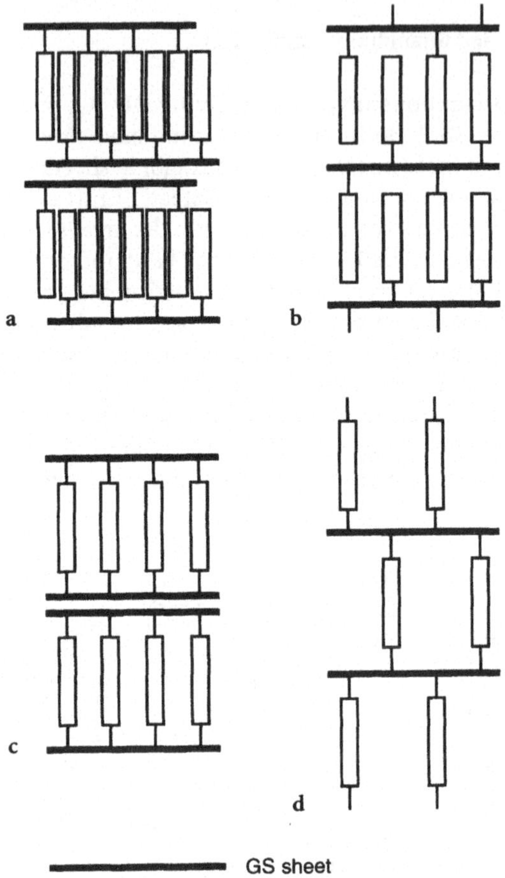

Fig. 17a–d Schematic representations of: **a** bilayer stacking in guanidinium sulfonate (GS) compounds; **b** continuous single layer stacking in GS compounds; **c** pillared bilayer stacking in guanidinium disulfonate compounds; **d** pillared brick stacking in guanidinium disulfonate compounds [138, 208]

to its ability to adopt either of these two stacking motifs and the inherent flexibility of the GS sheets, which are able to pucker in order to minimise void space. Compounds typically have one unit cell length of approximately 7.5 Å, which represents the S···S separation along a GS tape. A second unit cell length varies between 7.3 and 13 Å depending on the degree of puckering observed in the GS sheets, while the third depends on the separation between the sheets, which is dictated by the size of the substituents.

Competition arising from the inclusion of additional hydrogen bonding groups can disrupt the formation of the GS sheets and guanidinium salts of both 2,4,6-trinitrobenzenesulfonate and *p*-carboxybenzenesulfonate adopt more complex hydrogen-bonded structures, though retain GS hydrogen-bonded tapes [136, 138]. Inclusion of guest molecules can also change the structure adopted – guanidinium *p*-bromobenzenesulfonate crystallises in a bilayer structure, but inclusion of *p*-xylene or 2-chlorotoluene within the crystal structure leads to single layer stacking or tubular stacking respectively [139]. The use of guanidinium alkylbenzenesulfonates in forming smectic liquid crystals has been described, and the GS sheets shown by IR spectroscopy to persist in the disordered smectic phase [140].

The use of a disulfonate allows the GS hydrogen bonding sheets to be linked together, with the spacer group between the sulfonates acting as a pillar, and the length of this pillar determining the spacing between the sheets [141]. Both alkyl and aryl pillars can be incorporated, and chiral pillars based on tartrate have also been employed [142]. As with sulfonates there are two main modes of stacking possible – pillared bilayer (Fig. 17c) and pillared brick (or pillared continuous single-layer) (Fig. 17d) – though a range of other types of stacking have been identified [143]. These arise due to the ability of the GS host framework to respond to the size and shape of guest molecules, which have been demonstrated to template assembly formation. In many of these compounds the GS sheets exhibit a variation in the hydrogen bonding, with a shifted tape motif (Fig. 16b) adopted instead of the quasihexagonal array.

The type of stacking adopted depends on both the nature of the guest molecules and the size of the pillar. Thus guanidinium 4,4′-biphenyldisulfonate forms pillared bilayer structures with small guests such as toluene, styrene and *m*-xylene, but the more open and less dense pillared brick structures with larger guests such as 1,4-dibromobenzene and 1-nitronaphthalene [144]. The adoption of different structures for the same GS compound can be referred to as *architectural isomerism*. Guests that led to pillared brick structures with 4,4′-biphenyldisulfonate gave pillared bilayer structures with azobenzene-4,4′-disulfonate [145]. This can be rationalised on the basis of the longer length of the azobenzene-4,4′-disulfonate pillar, which leads to longer cavities in the bilayer structure.

Pillared brick structures can be induced by incorporation of guest molecules that are identical to the organic portion of the pillar. Thus, for example, guanidinium 2,6-anthracenedisulfonate crystallises with three guest molecules of anthracene to give a pillared brick structure in which the anthracene molecules and anthracenedisulfonate pillars pack in a near identical manner to that observed in the crystal structure of anthracene, with the disulfonates effectively replacing every fourth molecule in the herringbone motif of the pure guest [146]. Cooperative steric interactions between the pillars and the guests has been shown to influence ordering of guest molecules within the framework [147]. Co-crystallisation with guanidinium disulfonates has been used to separate mixtures of isomers, with the greatest selectivities occurring when the inclusion compounds are architectural isomers [148].

In a puckered pillared brick network, the puckering ensures that each GS sheet is polar, though as adjacent sheets tilt in opposite directions the overall framework is centric. Acentric polar versions of the pillared brick framework can be prepared using disulfonates in which the angle between the two C-S vectors is less than 180°. Such disulfonates ensure that the polarities of each GS sheet are orientated in the same direction, so the overall framework becomes polar. Mesitylenedisulfonate forms inclusion compounds with a range of guest molecules all of which exhibit the anticipated structures. Inclusion of a guest molecule

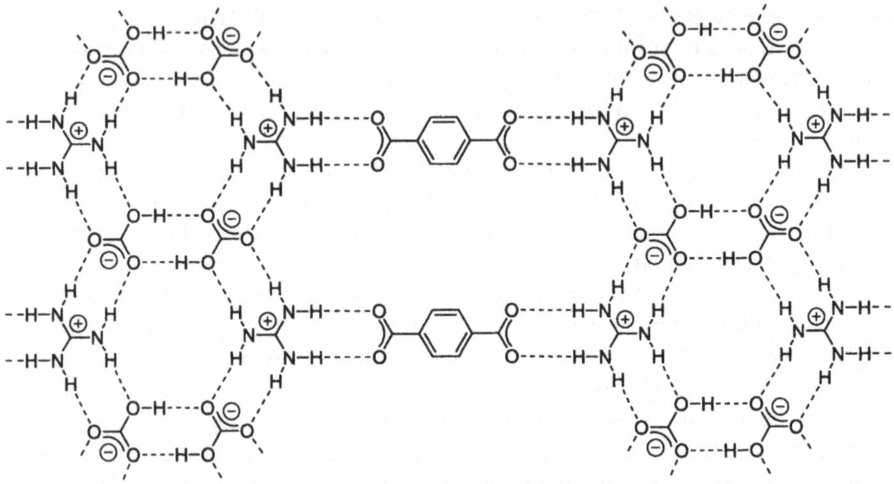

Fig. 18 Linking of guanidinium hydrogen carbonate tapes into sheets by hydrogen bonding to terephthalate anions [150]

that exhibits second harmonic generation activity leads to a GS framework in which that activity is maintained [149].

Replacement of the sulfonate with hydrogen carbonate can lead to tapes in which each hexagon consists of two guanidinium cations and four hydrogen carbonate anions. In the compound $(NBu_4)_3[C(NH_2)_3]_5(HCO_3)_4$[terephthalate]$_2$·$2H_2O$, terephthalate ions bridge guanidinium hydrogen carbonate tapes to give sheets (Fig. 18), which are interconnected via hydrogen bonding with additional guanidinium cations, which act as pillars [150].

4.3 Bis(Amidinium) Dicarboxylates

Bis(amidinium) dicarboxylates form tape structures in which each cation possesses two DD faces, and each anion two AA faces (Fig. 19) [151–153]. These tapes can be linked into sheets via the addition of extra dicarboxylic acid or by introduction of further hydrogen bonding groups into the cations [154] or the anions [155]. The same bis(amidinium) cations form sheet structures with $[Fe(CN)_6]^{3-}$,

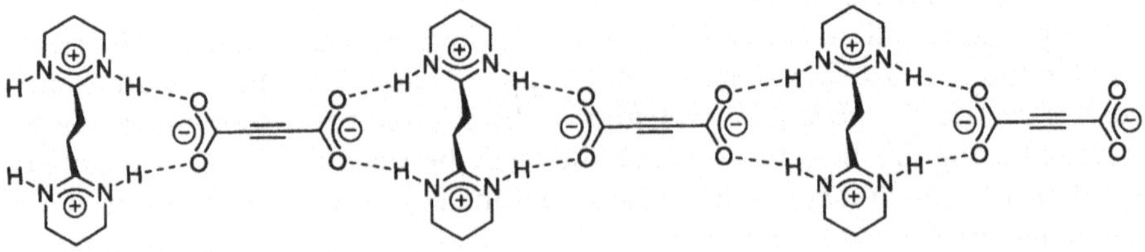

Fig. 19 Hydrogen-bonded tapes in the structure of a bis(amidinium) acetylenedicarboxylate compound [151]

with all NH groups hydrogen bonding to cyanide nitrogen atoms. Since each octahedral anion forms hydrogen bonds with three cations, each anion is in a chiral environment, although overall the network is achiral [156].

4.4 Thiourea Dicarboxylate Complexes and Related Systems

Both thiourea (tu) and thiosemicarbazide (tsc) are bifunctional ligands, containing a DD face in addition to one or two co-ordination sites. The reaction of $[Zn(tu)_4]^{2+}$ with a dicarboxylate normally occurs with displacement of thiourea to give coordination polymers of the type [Zn(tu)$_2$(μ-dicarboxylate)] in which the chains are cross-linked by DD-AA interactions [157]. The fumarate derivative contains identical inter-plane hydrogen bonding to that observed in $(NEt_4)_2$[fumarate]·2tu [158], in which the zinc atom has formally been replaced by two tetraethylammonium cations.

Bis(thiosemicarbazide)-nickel [159] and -zinc [160] dications crystallise with dicarboxylates to give hydrogen-bonded tapes in the solid state. Since the cations and anions both contain additional hydrogen bonding groups to those involved in tape formation, the manner in which the tapes are linked together is determined by the substitution pattern within the thiosemicarbazide. Thus for the series of nickel compounds *trans*-[NiL$_2$][terephthalate] [161], when L is $NH_2C(S)NHNH_2$ the tapes are cross-linked through hydrogen bonds involving a thioamido NH proton, whereas when L is NHMeC(S)NHNH$_2$, this proton has been substituted so sheet formation occurs through hydrogen bonds involving the amino NH proton. In the case in which L is NHMeC(S)NHNMe$_2$ (tmtsc), all the thiosemicarbazide protons not involved in tape formation have been substituted and sheet formation occurs through hydrogen bonds involving OH protons from water molecules co-ordinated in the axial positions (Fig. 20). In cases containing mutually *cis* thiosemicarbazide ligands, there is a second opportunity for tape formation to arise involving $R_2^2(8)$ motifs, in which parallel NH groups on the two amino groups act as donors. In the structure of *cis*-[Ni{NHEtC(S)NHNH$_2$}$_2$][terephthalate] it is this arrangement that is adopted in the solid state [162].

Addition of a dicarboxylate to $[Zn(tmtsc)_2]^{2+}$ leads to a variety of structural types, with the major factor determining the structure adopted being the relative orientation of the carboxylate groups in the anion. Linear dicarboxylates such as fumarate and terephthalate gives hydrogen-bonded tape structures, though the terephthalate compound contains 'expanded' dimeric cations in the observed product [{Zn(tmtsc)(OH$_2$)}$_2$(μ-terephthalate)]terephthalate·2H$_2$O **30** [163].

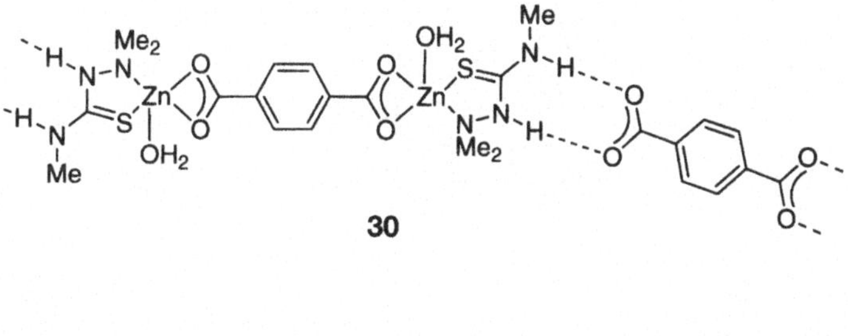

30

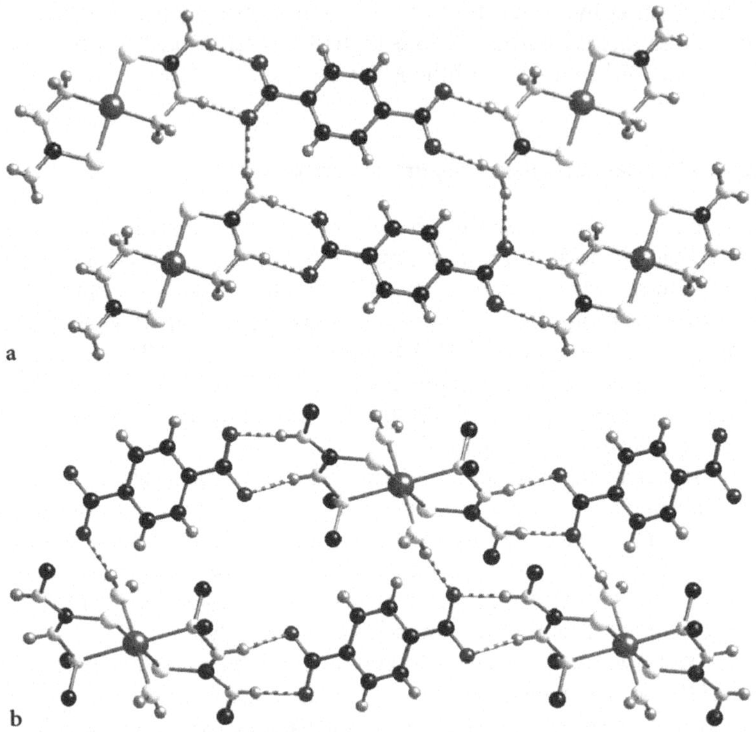

Fig. 20 **a** Cross-linking of $[Ni(tsc)_2]^{2+}$ terephthalate tapes through NH···O hydrogen bonds. **b** Cross-linking of $[Ni(tmtsc)_2(OH_2)_2]^{2+}$ terephthalate tapes through OH···O hydrogen bonds [161]

Compounds containing amidino-*O*-alkylurea ligands also contain DD faces. Recrystallisation of (bis(amidino-*O*-ethylurea)ethane)copper tetrafluoroborate **31** from methanol afforded a structure in which the difluorodimethoxyborate anion, generated by methanolysis, is trapped in tapes. Each anion presents two AA faces, each based on one oxygen and one fluorine atom [164]. As with the DA-AD interactions described earlier, DD-AA interactions can involve coordinated water ligands as donor groups. In the structure of a hydrated hexaaquacobalt(II) tetra(carboxyl)tetrathiafulvalene, DD-AA interactions were observed with mutually *cis* waters acting as the hydrogen bond donors. This compound is noteworthy as it undergoes desolvation without loss of monocrystallinity, and the unsolvated compound contains a different hydrogen bonding pattern [165].

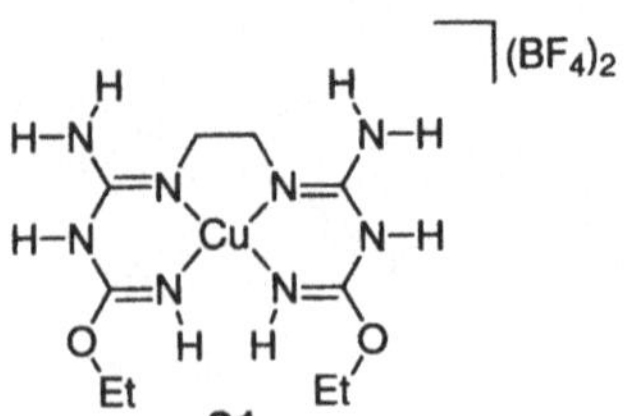

31

5
Systems Based on ADA-DAD Interactions

Much of the important early work in organic crystal engineering was undertaken using ADA-DAD hydrogen-bonded systems. The archetypal structure is that of the melamine-cyanuric acid adduct, which has a hexagonal sheet structure (Fig. 21). Although the inspiration for many studies, the sheet structure of this adduct was only confirmed crystallographically in 1999, with the problem of insolubility being surmounted using hydrothermal methods [166].

The groups of Lehn [167] and Whitesides [168] have studied the effects of substituents in the melamine and cyanuric acid components on the manner in which the molecules interact. Both groups used barbituric acids as components containing two ADA faces, whereas for the components bearing two DAD faces Lehn's group used 2,4,6-triaminopyrimidines and Whiteside's group *N,N'*-disubstituted melamines. The size of the substituents on the melamine has been shown to be an important factor in determining the type of structure adopted. Melamines with sterically undemanding substituents such as 4-chlorophenyl gave linear tapes when crystallised with 5,5-diethylbarbituric acid [169], whereas more demanding groups such as *tert*-butyl led to crinkled tapes in order to avoid unfavourable interactions between the butyl groups (Fig. 22) [170]. Very demanding groups such as 4-(*tert*-butyl)phenyl gave discrete [3+3] rosettes instead of extended structures [171]. Incorporation of phenyl groups substituted in the 3-position into melamines allows for a greater number of molecular conformations, and consequently co-crystals of these compounds with 5,5-diethylbarbituric acid show a greater variety of packing, with both linear and crinkled tapes observed, and a higher frequency of solvent inclusion noted [172]. *N,N'*-Diphenyl melamine gave a linear tape structure with 5,5-diethylbarbituric acid but, perhaps surprisingly, a crinkled tape structure with 5,5-dimethylbarbituric

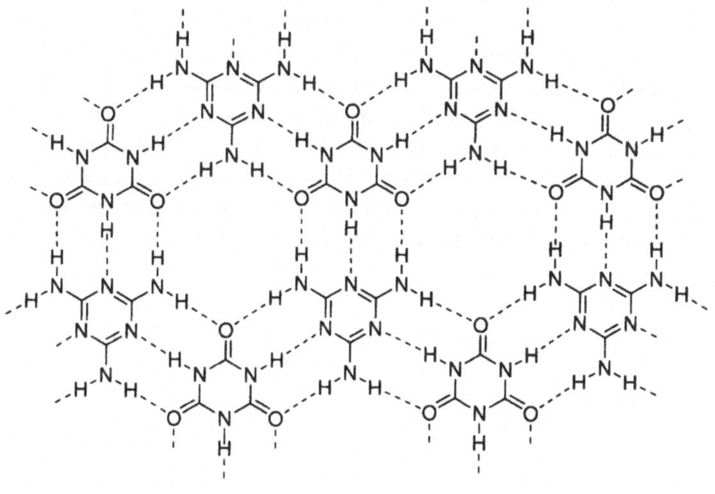

Fig. 21 The hexagonal melamine-cyanuric acid hydrogen-bonded array [166]

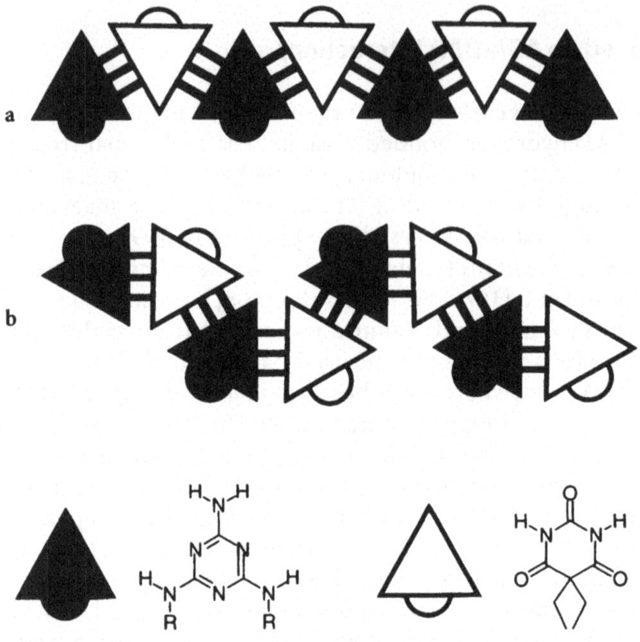

Fig. 22a, b Schematic representations of: **a** linear tapes; **b** crinkled tapes observed in structures based on *N,N'*-disubstituted melamines and 5,5-diethylbarbituric acid [169]

acid.[173] Adoption of this structure allows one of the phenyl substituents to engage in $\pi\cdots\pi$ interactions.

Judicious choices of substituents have led to adducts that possess a wide range of properties, and examples include monolayers on gold surfaces [174] or at air-water interfaces [175–178] and for use as liquid crystals [179]. Expanded components, in which the distance between the faces is increased have also been studied [180, 181]. TEM studies have shown the existence of strands, whose diameter is consistent with either stacked rosettes or helical tapes. TEM has also been used to study the products from reactions between bismelamines and biscyanuric acids [182], and bis(diamidopyridines) and bisuracils [183].

Co-crystallisation of the 2,4-diamino-6-alkyltriazine **32** with uracil derivative **33** gave 1:1 co-crystals in which ADA-DAD pairs are linked into tapes through DA-AD interactions involving the uracils. Incorporation of hydroxymethyl groups into the 6-positions of the triazine and uracil tectons led to the tapes

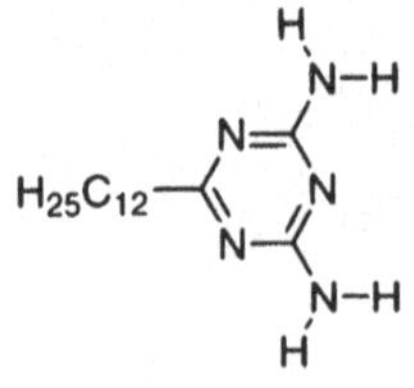

32

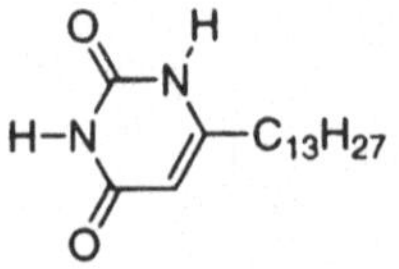

33

being connected into sheets through additional hydrogen bonds [184]. Co-crystallisation of melamine with succinimide led to a sheet structure in which all of the hydrogen bond donors and acceptors were satisfied. In contrast, co-crystallisation of melamine with glutarimide gave a tape structure in which melamine·2glutarimide units are connected by two NH···O hydrogen bonds [185]. Co-crystallisation of 2,4-diamino-6-phenyl-1,3,5-triazine (dpt) with phthalimide gave a 2:1 adduct whose structure consists of tapes formed from DA-AD interactions between dpt molecules, with every other dpt forming a ADA-DAD interaction with a phthalimide molecule [186]. Incorporation of the ADA face into a five-membered ring leads to shorter hydrogen bonds involving the central NH group than when the face is part of a six-membered ring.

Either DAD or ADA faces can be included into a bifunctional ligand, allowing for the incorporation of a metal centre into the structure [23]. Typically the resultant structures are more complex than those arising from organic systems due in part to the decreased symmetry of the metal complex. Mingos and co-workers have used dithiobiureto ligands to form a nickel complex [Ni(dtb)$_2$] **34** that bears two DAD hydrogen bonding faces. This compound co-crystallises with uracil to give a sheet structure [187], with bemegride **35** to give a tape structure and with 1,8-naphthalimide **36** to give discrete [1+2] units [188]. The dimensionality of the product is influenced by the presence or absence of additional hydrogen bonding groups and the steric demands of the molecules. [Ni(dtb)$_2$]·2(1,8-naphthalimide) has the appropriate hydrogen bond donor and acceptor groups to form a tape structure, but is prevented from doing so by the size of the naphthalene moiety. Incorporation of melamine to form the macrocyclic ligand **37** has led to copper complexes containing DAD faces. The crystal structure of [Cu(**37**)](ClO$_4$)$_2$·H$_2$O reveals the presence of dimers in which the molecules are linked by $R_2^2(8)$ motifs, with these units further linked into tapes via hydrogen bonds involving the included water molecule. Crystallisation of this complex with cyanuric acid gave [Cu(**37**)(cyanurate)]ClO$_4$·3H$_2$O in which ligands containing both ADA and DAD groups are coordinated. This complex is consequently self-complementary, and tapes of cations are observed in the crystal structure [189].

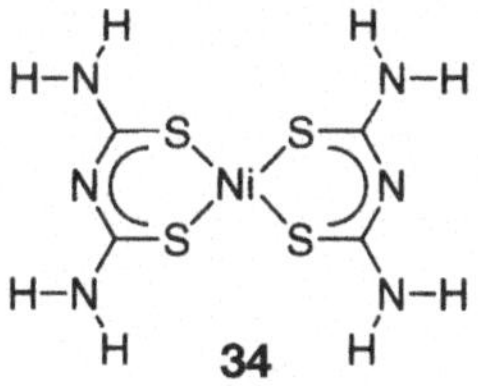

34

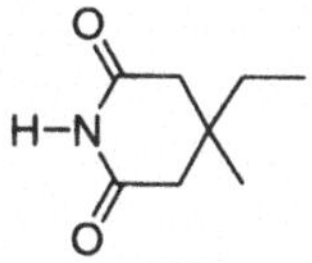

35

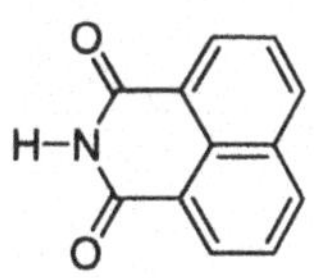

36

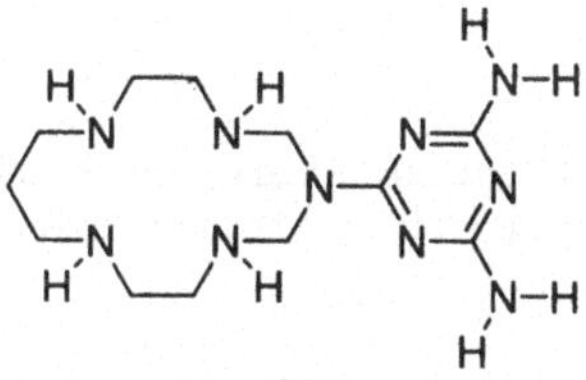

37

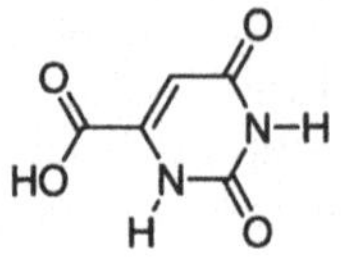

38

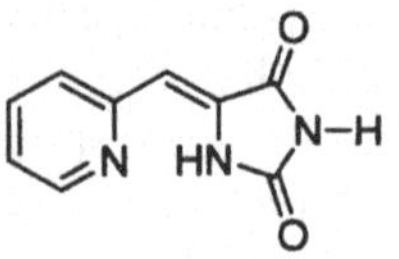

39

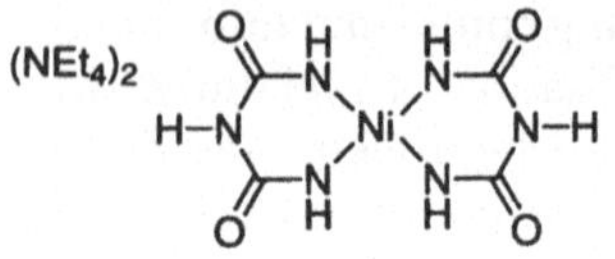

40

Bifunctional ligands employing ADA faces have also attracted attention. Deprotonation of orotic acid (H_2orot) **38** [190] and 5-(2-pyridylmethylene)hydantoin (Hpyhy) **39** [191] leads to a dianionic *N,O*-donor ligand and a monoanionic *N,N*-donor ligand respectively. The steric demands of any co-ligands are critical in determining whether an extended or a discrete structure is formed. For example, the co-crystal of NBu_4[Rh(cod)(orot)] and 2,6-diaminopyridine forms a sheet structure in which the principal motif is the ADA-DAD interaction between the orotate and the 2,6-diaminopyridine, and further hydrogen bonds connect these [1+1] units together [192]. In contrast, the co-crystal of [Pt(dppe)(orot)] and 2,6-diaminopyridine forms similar [1+1] units, though further hydrogen bonds lead only to dimers of these, with further extension of the network prevented by the steric bulk of the dppe ligands [190]. The complex $[Cu(pyhy)_2]$ co-crystallises with melamine to give corrugated sheets in which $[Cu(pyhy)_2]$·2melamine units are linked by pairs of $R_2^2(8)$ motifs between adjacent melamine molecules (Fig. 23) [193]. The nickel biureto complex $(NEt_4)_2[Ni(bu)_2]$ **40** forms 1:2 co-crystals with 2,4-diamino-6-phenyl-1,3,5-triazine (dpt) in which the $[Ni(bu)_2]^{2-}$·2dpt units are linked into tapes through additional NH···O hydrogen bonds [186].

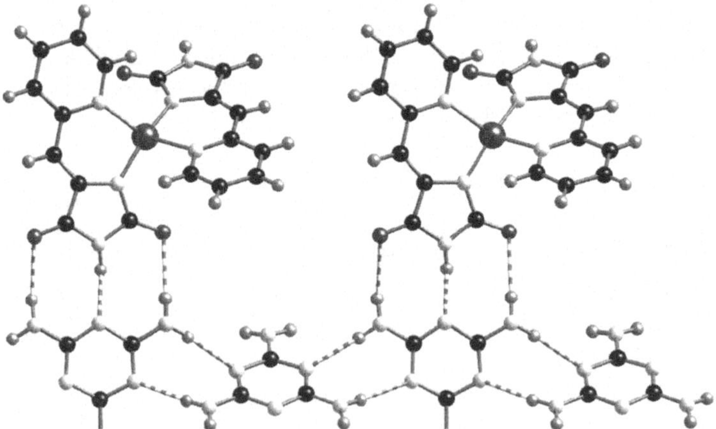

Fig. 23 Part of the structure of $[Cu(pyhy)_2]$·2melamine, showing ADA-DAD interactions between pyhy ligands and melamines, and DA-AD interactions between melamine molecules [193]

6
Systems Based on DDA-AAD, DDA′-A′DD and DDD-AAA Interactions

Despite being theoretically more appealing than ADA-DAD systems due to the favourable secondary interactions, DDA-AAD and DDD-AAA systems have received considerably less attention. One reason for this is the lack of suitable precursor compounds. The pyrimidinone compound **41** contains one DDA face and one AAD face, so is self-complementary. Since a carbonyl group acts as a hydrogen bond acceptor on both faces, tape formation dictates that each molecule is orientated in an opposite direction from its neighbours. The same logic also applies when an NH_2 group acts as a hydrogen bond donor to two faces. As expected, the crystal structure of **41** contains tapes with the molecules linked by DDA-AAD interactions [194]. The pyrido[4,3-*d*]pyrimidine **42** also contains DDA and AAD faces though in this case no groups are shared between the faces, and the relative orientation of these is such that the molecules give discrete cyclic hexamers as opposed to infinite tapes [195].

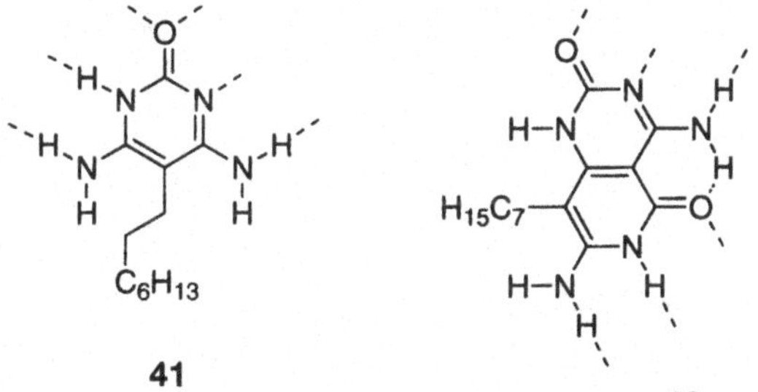

41 **42**

Although not normally considered self-complementary, DDA faces can hydrogen bond together provided the acceptor is a carbonyl oxygen atom, so able to form hydrogen bonds with both donors in a $R^1_2(6)$ motif, similar to that observed in ureas. This interaction can be denoted DDA′-A′DD. Oxalurate complexes, for example, form solid state structures in which pairs of ligands are connected by four hydrogen bonds between DDA faces in this manner. This, in combination with DA-AD interactions, gives a sheet structure containing pores that are filled by included and coordinated water molecules (Fig. 24) [196].

In situations where cyanuric acid groups are sufficiently acidic, and their counterparts sufficiently basic, proton transfer might be expected, which would result in the formation of ion-pair reinforced tapes in which half of the ADA-DAD interactions have been converted to DDD-AAA interactions. This has been observed in the products from the co-crystallisations of *N*-(3-hydroxylpropyl)cyanuric acid with 5-butyl-2,4,6-triaminopyrimidine [197] and 1-(4-carboxybutyl)-1,3,5-triazine-2,4,6-trione with 5-(2-aminoethyl)-2,4,6-triaminopyrimidine [198]. In the latter example, a second proton transfer occurs from the carboxylic acid group to the amino group, and hydrogen bonding between the resultant ammonium and carboxylate groups links the tapes into sheets (Fig. 25).

Although the triple hydrogen bond systems that have been addressed so far in this review contain a linear arrangement of the hydrogen bonding groups, this is not necessarily the case. The functionalised *cis,cis*-cyclohexane-1,3,5-tricarbox-

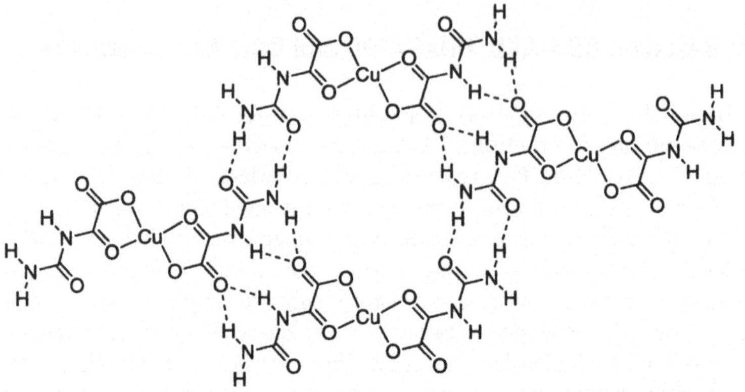

Fig. 24 Hydrogen-bonded sheets in $[Co(oxalurate)_2(OH_2)_2]\cdot 2H_2O$, with water molecules omitted for clarity [196]

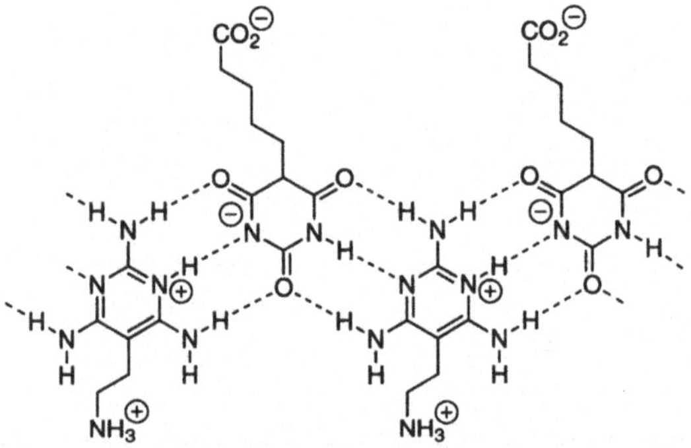

Fig. 25 DDD-AAA and ADA-DAD interactions in tapes formed from 1-(4-carboxybutyl)-1,3,5-triazine-2,4,6-trione with 5-(2-aminoethyl)-2,4,6-triaminopyrimidine [198]

amide **43** self-assembles with three NH···O hydrogen bonds linking the molecules into rods [199]. In addition, the interaction of an ammonium ion (or ammine ligand) with a [18]crown-6 crown ether involves three NH···O hydrogen bonds. Since the NH groups are directed at the oxygen atoms the association energy is greater than it would be using, for example, [15]-crown-5. However, this crown ether has been employed to form hydrogen-bonded structures, as in the structure of $[UO_2Cl_2(H_2O)_3]\cdot$[15]-crown-5 which consists of chains connected by, and linked through, OH···O hydrogen bonds [200].

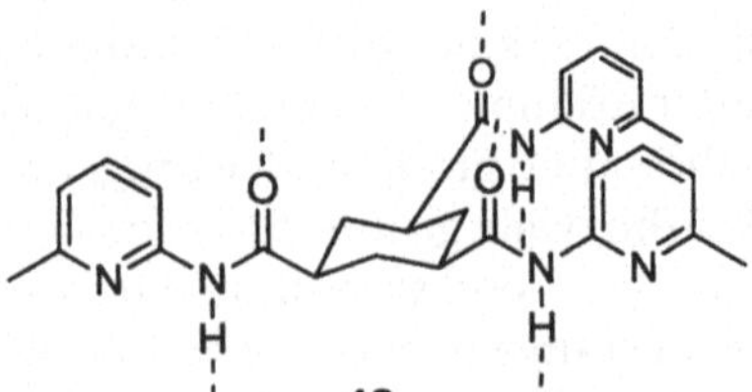

43

7
Systems in Which Molecules Are Linked by Four or More Hydrogen Bonds

Both ADAD-DADA and AADD-DDAA systems have attracted increasing attention recently [201]. As both the ADAD and AADD faces are self-complementary, quadruple hydrogen bonds can occur in one-component systems. Acylation of diaminotriazines leads to compounds such as **44**. In the crystal structure of **44**, quadruply hydrogen-bonded dimers are linked via $R_2^2(8)$ motifs into tapes [202]. Molecules containing two diaminotriazines form tapes with all molecules linked by ADAD-DADA interactions [203]. ADAD-DADA interactions have also been observed in inorganic systems, for example between the nitrilotri(methylphosphonate) ligands in $[Mn(H_2O)_3\{HN(CH_2PO_3H)_3\}]$ [204].

The self-assembly of molecules containing two AADD faces into supramolecular polymers has been observed by employing molecules containing two ureidopyrimidone moieties such as **45** [27], and the same logic has been used to construct tapes in which fullerenes [205] and siloxanes [206] are present as part of the backbone.

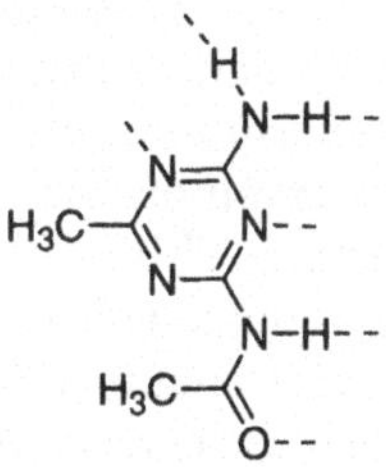

44

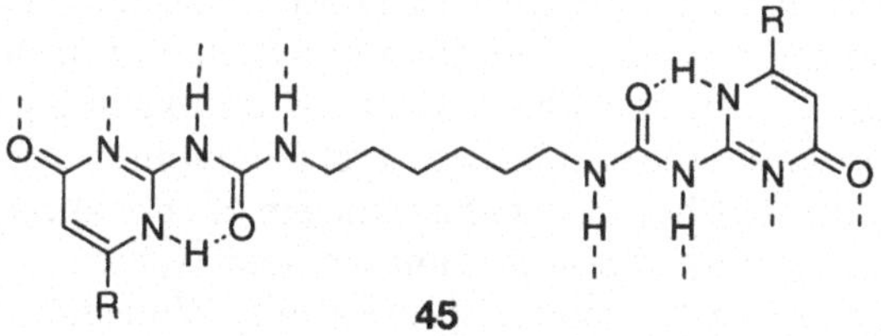

45

Extension of the ADA-DAD motif by incorporating two sets of pairs of ADA or DAD faces leads to the potential for forming sextuple hydrogen bonds. The molecules **46** and **47** have been observed to interact strongly by NMR spectroscopy to give linear tapes [207].

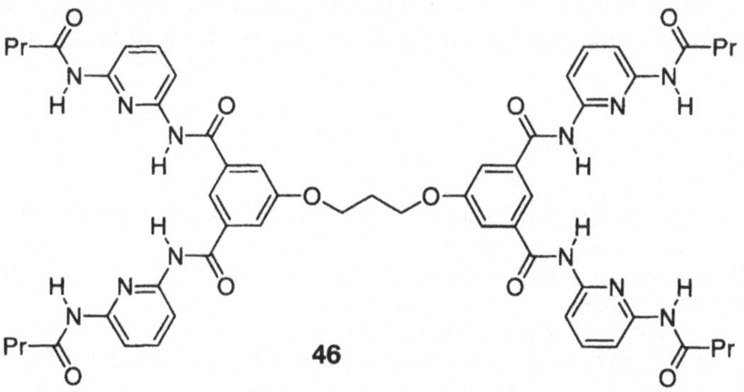

46

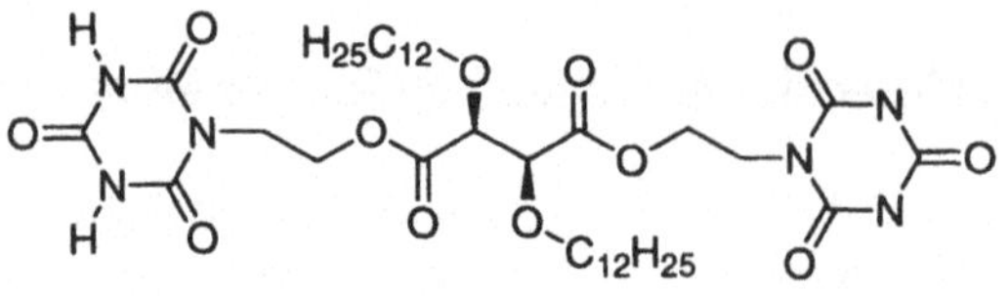

47

8
Outlook

Crystal engineering is today a vibrant and multidisciplinary area of science. The strength of the area is witnessed by the recent inauguration of several international journals in which the topic plays a major role, including *Crystal Engineering* (Pergamon-Elsevier, 1998-), *CrystEngComm* (RSC, 1999-) and *Crystal Growth and Design* (ACS, 2001-). Although the routine preparation of compounds with pre-defined extended structure and function remains a distant goal, vast progress has been made, especially with regard to defining the limits of the various approaches, and in certain systems extended structures are now predictable with a reasonable degree of certainty. Crystal engineering has so far concentrated rather more on structure than function, both in terms of determining the extended structures of new compounds and exploiting the vast reservoir of data contained within the Cambridge Crystallographic Structural Database. This focus on structure has been used as a criticism of crystal engineering, though without an understanding of how to control assembly, the synthesis of functional materials becomes largely a matter of serendipity. While not yet complete, the crystal engineering toolkit is now large enough for it to be useful, and undoubtedly research in the next 10–15 years will focus more and more on the preparation of structures with specific functions in mind. Porous networks are already attracting considerable attention as 'synthetic zeolites', with potential uses in catalysis, gas storage and separations. Materials with non-linear optical, magnetic and coordinating properties have also attracted interest, and the ideas behind crystal engineering are being used in the generation of self-assembled monolayers and liquid crystals and in the study of biomineralisation. It is anticipated that all of these applications, and others besides, will build upon the design strategies outlined in this review.

9
References

1. Atwood JL, Davies JED, MacNicol DD, Vögtle F (eds) (1996) Comprehensive supramolecular chemistry. Pergamon, Oxford
2. Braga D, Desiraju GR, Miller JS, Orpen AG, Price SL (2002) Cryst Eng Comm 4:500
3. Braga D (2000) J Chem Soc Dalton Trans:3705
4. Schmidt GMJ (1971) Pure Appl Chem 27:647
5. Desiraju GM (1989) Crystal engineering, the design of organic solids. Elsevier, Amsterdam
6. Motherwell WDS, Ammon HL, Dunitz JD, Dzyabchenko A, Erk P, Gavezzotti A, Hofmann DWM, Leusen FJJ, Lommerse JPM, Mooij WTM, Price SL, Scheraga H, Schweizer B, Schmidt MU, van Eijck BP, Verwer P, Williams DE (2002) Acta Crystallogr Sect B 58:647

7. Meléndez RE, Hamilton AD (1998) Top Curr Chem 198:97
8. Aakeröy CB, Beatty AM (2001) Aust J Chem 54:409
9. Blake AJ, Champness NR, Hubberstey P, Li W-S, Withersby MA, Schröder M (1999) Coord Chem Rev 183:117
10. Robson R (2000) J Chem Soc, Dalton Trans:3735
11. Eddaoudi M, Moler DB, Li H, Chen B, Reineke TM, O'Keeffe M, Yaghi OM (2001) Acc Chem Res 34:319
12. Jeffrey GA (1997) An introduction to hydrogen bonding. OUP, New York
13. Steiner T (2002) Angew Chem Int Ed 41:48
14. Desiraju GR (2002) Acc Chem Res 35:565
15. Gillon AL, Lewis GR, Orpen AG, Rotter S, Starbuck J, Wang X-M, Rodríguez-Martín Y, Ruiz-Pérez C (2000) J Chem Soc Dalton Trans:3897
16. Desiraju GR (1995) Angew Chem Int Ed Engl 34:2311
17. Simard M, Su D, Wuest JD (1991) J Am Chem Soc 113:4696
18. Marsh A, Nolen EG, Gardinier KM, Lehn J-M (1994) Tetrahedron Lett 35:397
19. Etter MC (1991) J Phys Chem 95:4601
20. Bernstein J, Davis RE, Shimoni L, Chang N-L (1995) Angew Chem Int Ed Engl 34:1555
21. Jorgensen WL, Severance DL (1991) J Am Chem Soc 113:209
22. Pranata J, Wierschke SG, Jorgensen WL (1991) J Am Chem Soc 113:2810
23. Burrows AD, Chan C-W, Chowdhry MM, McGrady JE, Mingos DMP (1995) Chem Soc Rev 24:329
24. MacDonald JC, Whitesides GM (1994) Chem Rev 94:2383
25. Nguyen TL, Fowler FW, Lauher JW (2001) J Am Chem Soc 123:11,057
26. Burrows AD (2002) Sci Prog 85:199
27. Sijbesma RP, Beijer FH, Brunsveld L, Folmer BJB, Hirschberg JHKK, Lange RFM, Lowe JKL, Meijer EW (1997) Science 278:1601
28. Holman KT, Pivovar AM, Swift JA, Ward MD (2001) Acc Chem Res 34:107
29. Duchamp DJ, Marsh RE (1969) Acta Crystallogr Sect B 25:5
30. Herbstein FH, Kapon M, Reisner GM (1987) J Inclusion Phenom 5:211
31. Kolotuchin SV, Fenlon EE, Wilson SR, Loweth CJ, Zimmerman SC (1995) Angew Chem Int Ed Engl 34:2654
32. Kolotuchin SV, Thiessen PA, Fenlon EE, Wilson SR, Loweth CJ, Zimmerman SC (1999) Chem Eur J 5:2537
33. Dunitz JD (1998) Chem Eur J 4:745
34. Ermer O, Eling A (1988) Angew Chem Int Ed Engl 27:829
35. Ermer O (1988) J Am Chem Soc 110:3747
36. Ermer O, Lindenberg L (1991) Helv Chim Acta 74:825
37. Holy P, Závada J, Císarová I, Podlaha J (1999) Angew Chem Int Ed 38:381
38. Diskin-Posner Y, Kumar RK, Goldberg I (1999) New J Chem 23:885
39. Lin Q, Geib SJ, Hamilton AD (1998) J Chem Soc Perkin Trans 2:2109
40. Ashton PR, Fyfe MCT, Hickingbottom SK, Menzer S, Stoddart JF, White AJP, Williams DJ (1998) Chem Eur J 4:577
41. Braga D, Grepioni F, Sabatino P, Desiraju GR (1994) Organometallics 13:3532
42. Edwards DA, Mahon MF, Paget TJ (1998) Polyhedron 17:4121
43. Qin Z, Jennings MC, Puddephatt RJ, Muir KW (2002) Inorg Chem 41:5174
44. Gianneschi NC, Tiekink ERT, Rendina LM (2000) J Am Chem Soc 122:8474
45. Sekiya R, Nishikiori S (2002) Chem Eur J 8:4803
46. Aakeröy CB, Beatty AM, Lorimer KR (2000) J Chem Soc Dalton Trans:3869
47. Takusagawa F, Koetzle TF (1979) Acta Crystallogr Sect B 35:2888
48. Braga D, Maini L, Grepioni F, Elschenbroich C, Paganelli F, Schiemann O (2001) Organometallics 20:1875
49. Braga D, Maini L, Polito M, Rossini M, Grepioni F (2000) Chem Eur J 6:4227
50. Braga D, Maini L, Paganelli F, Tagliavini E, Casolari S, Grepioni F (2001) J Organomet Chem 637:609
51. Brammer L, Rivas JCM, Atencio R, Fang SY, Pigge FC (2000) J Chem Soc Dalton Trans:3855

52. Fraser CSA, Jenkins HA, Jennings MC, Puddephatt RJ (2000) Organometallics 19:1635
53. Allen FH, Motherwell WDS, Raithby PR, Shields GP, Taylor R (1999) New J Chem 23:25
54. Kitaigorodskii AI (1961) Organic chemical crystallography. Consultants Bureau, New York
55. Perlstein J, Steppe K, Vaday S, Ndip EMN (1996) J Am Chem Soc 118:8433
56. Desmartin PG, Williams AF, Bernardinelli G (1995) New J Chem 19:1109
57. Frankenbach GM, Etter MC (1992) Chem Mater 4:272
58. Sharma CVK, Clearfield A (2000) J Am Chem Soc 122:4394
59. Mao J-G, Wang Z, Clearfield A (2002) Inorg Chem 41:3713
60. Ma B-Q, Gao S, Sun H-L, Xu G-X (2001) J Chem Soc Dalton Trans:130
61. Leiserowitz L, Schmidt GMJ (1969) J Chem Soc A:2372
62. Luo T-JM, Palmore GTR (2000) J Phys Org Chem 13:870
63. Gong B, Zheng C, Skrzypczak-Jankun E, Yan Y, Zhang J (1998) J Am Chem Soc 120:11194
64. Kuduva SS, Bläser D, Boese R, Desiraju GR (2001) J Org Chem 66:1621
65. Palacin S, Chin DN, Simanek EE, MacDonald JC, Whitesides GM, McBride MT, Palmore GTR (1997) J Am Chem Soc 119:11,807
66. Chin DN, Palmore GTR, Whitesides GM (1999) J Am Chem Soc 121:2115
67. Williams LJ, Jagadish B, Lyon SR, Kloster RA, Carducci MD, Mash EA (1999) Tetrahedron 55:14281
68. Williams LJ, Jagadish B, Lansdown MG, Carducci MD, Mash EA (1999) Tetrahedron 55:14301
69. Palmore GTR, McBride MT (1998) Chem Commun:145
70. Luo T-JM, Palmore GTR (2002) Cryst Growth Des 2:337
71. McBride MT, Luo T-JM, Palmore GTR (2001) Cryst Growth Des 1:39
72. Johnson DW, Palmer LC, Hof F, Iovine PM, Rebek J Jr (2002) Chem Commun 2228
73. Wu A, Fettinger JC, Isaacs L (2002) Tetrahedron 58:9769
74. Aakeröy CB, Hughes DP, Nieuwenhuyzen M (1996) J Am Chem Soc 118:10,134
75. Wang X, Simard M, Wuest JD (1994) J Am Chem Soc 116:12,119
76. Aakeröy CB, Beatty AM, Zou M (1998) Cryst Eng 1:225
77. Aakeröy CB, Beatty AM, Nieuwenhuyzen M, Zou M (2000) Tetrahedron 56:6693
78. Batchelor E, Klinowski J, Jones W (2000) J Mater Chem 10:839
79. Edwards MR, Jones W, Motherwell WDS (2002) Cryst Eng 5:25
80. Etter MC, Reutzel SM (1991) J Am Chem Soc 113:2586
81. Etter MC, Urbanczyk-Lipkowska Z, Zia-Ebrahimi M, Panunto TW (1990) J Am Chem Soc 112:8415
82. Hollingsworth MD, Brown ME, Santarsiero BD, Huffman JC, Goss CR (1994) Chem Mater 6:1227
83. Coe S, Kane JJ, Nguyen TL, Toledo LM, Wininger E, Fowler FW, Lauher JW (1997) J Am Chem Soc 119:86
84. Pedireddi VR, Belhekar D (2002) Tetrahedron 58:2937
85. Falvello LR, Pascual I, Tomas M, Urriolabeitia EP (1997) J Am Chem Soc 119:11,894
86. Hunks WJ, Jennings MC, Puddephatt RJ (2002) Inorg Chem 41:4590
87. Aakeröy CB, Beatty AM (1998) Cryst Eng 1:39
88. Chen C-H, Cai J, Feng X-L, Chen X-M (2002) Polyhedron 21:689
89. Kuehl CJ, Tabellion FM, Arif AM, Stang PJ (2001) Organometallics 20:1956
90. Qin Z, Jennings MC, Puddephatt RJ (2001) Inorg Chem 40:6220
91. Kurth DG, Fromm KM, Lehn J-M (2001) Eur J Inorg Chem:1523
92. Dai J, Yamamoto M, Kuroda-Sowa T, Maekawa M, Suenaga Y, Munakata M (1997) Inorg Chem 36:2688
93. Scherer M, Sessler JL, Moini M, Gebauer A, Lynch V (1998) Chem Eur J 4:152
94. Marsman AW, Leussink ED, Zwikker JW, Jenneskens LW, Smeets WJJ, Veldman N, Spek AL (1999) Chem Mater 11:1484
95. Marsman AW, van Walree CA, Havenith RWA, Jenneskens LW, Lutz M, Spek AL, Lutz ETG, van der Maas JH (2000) J Chem Soc Perkin Trans 2:501
96. Aakeröy CB, Beatty AM, Leinen DS (1998) J Am Chem Soc 120:7383
97. Aakeröy CB, Beatty AM, Leinen DS (2002) Cryst Eng Comm 4:310

98. Mazik M, Bläser D, Boese R (1999) Tetrahedron Lett 40:4783
99. Aakeröy CB, Beatty AM, Leinen DS (2001) Cryst Growth Des 1:47
100. Bensemann I, Gdaniec M, Polonski T (2002) New J Chem 26:448
101. Scheinbeim J, Schempp E (1976) Acta Crystallogr Sect B 32:607
102. Krische MJ, Lehn J-M, Kyritsakas N, Fischer J, Wegelius EK, Nissinen MJ, Rissanen K (1998) Helv Chim Acta 81:1921
103. Krische MJ, Lehn J-M, Kyritsakas N, Fischer J (1998) Helv Chim Acta 81:1909
104. Krische MJ, Lehn J-M, Cheung E, Vaughn G, Krische AL (1999) C R Acad Sci Ser II C 2:549
105. Krische MJ, Lehn J-M, Kyritsakas N, Fischer J, Wegelius EK, Rissanen K (2000) Tetrahedron 56:6701
106. Ziener U, Breuning E, Lehn J-M, Wegelius E, Rissanen K, Baum G, Fenske D, Vaughan G (2000) Chem Eur J 6:4132
107. Breuning E, Ziener U, Lehn J-M, Wegelius E, Rissanen K (2001) Eur J Inorg Chem 1515
108. Cromer DT, Ryan RR, Storm CB (1987) Acta Crystallogr Sect C 43:1435
109. Allen WE, Fowler CJ, Lynch VM, Sessler JL (2001) Chem Eur J 7:721
110. Tadokoro M, Nakasuji K (2000) Coord Chem Rev 198:205
111. Tadokoro M, Kanno H, Kitajima T, Shimada-Umemoto H, Nakanishi N, Isobe K, Nakasuji K (2002) Proc Natl Acad Sci USA 99:4950
112. Tadokoro M, Isobe K, Uekusa H, Ohashi Y, Toyoda J, Tashiro K, Nakasuji K (1999) Angew Chem Int Ed 38:95
113. Tadokoro M, Shiomi T, Isobe K, Nakasuji K (2001) Inorg Chem 40:5476
114. Öhrström L, Larsson K, Borg S, Norberg ST (2001) Chem Eur J 7:4805
115. Tadokoro M, Shiomi T, Shiromizu T, Isobe K, Matsumoto K, Nakasuji K (1997) Mol Cryst Liq Cryst A 306:235
116. Brunet P, Simard M, Wuest JD (1997) J Am Chem Soc 119:2737
117. Dahal S, Goldberg I (2000) J Phys Org Chem 13:382
118. Chan C-W, Mingos DMP, White AJP, Williams DJ (1996) Polyhedron 15:1753
119. Batsanov AS, Begley MJ, Hubberstey P, Stroud J (1996) J Chem Soc Dalton Trans 1947
120. Batsanov AS, Begley MJ, George MW, Hubberstey P, Munakata M, Russell CE, Walton PH (1999) J Chem Soc, Dalton Trans:4251
121. Burrows AD, Harrington RW, Mahon MF, Teat SJ (2002) Cryst Eng Comm 4:539
122. Garcia-Tellado F, Geib SJ, Goswami S, Hamilton AD (1991) J Am Chem Soc 113:9265
123. Fan E, Vicent C, Geib SJ, Hamilton AD (1994) Chem Mater 6:1113
124. Shan N, Bond AD, Jones W (2002) Tetrahedron Lett 43:3101
125. Simanek EE, Tsoi A, Wang CCC, Whitesides GM, McBride MT, Palmore GTR (1997) Chem Mater 9:1954
126. Vishweshwar P, Thaimattam R, Jaskolski M, Desiraju GR (2002) Chem Commun 1830
127. Vishweshwar P, Nangia A, Lynch VM (2002) J Org Chem 67:556
128. Hanessian S, Gomtsyan A, Simard M, Roelens S (1994) J Am Chem Soc 116:4495
129. Hanessian S, Saladino R, Margarita R, Simard M (1999) Chem Eur J 5:2169
130. Podesta TJ, Orpen AG (2002) Cryst Eng Comm 4:336
131. Zaman MB, Tomura M, Yamashita Y (2001) J Org Chem 66:5987
132. Zafar A, Geib SJ, Hamuro Y, Hamilton AD (1998) New J Chem 22:137
133. Zafar A, Melendez R, Geib SJ, Hamilton AD (2002) Tetrahedron 58:683
134. Papoutsakis D, Kirby JP, Jackson JE, Nocera DG (1999) Chem Eur J 5:1474
135. Katrusiak A, Szafranski M (1994) Acta Crystallogr Sect C 50:1161
136. Russell VA, Etter MC, Ward MD (1994) Chem Mater 6:1206
137. Russell VA, Etter MC, Ward MD (1994) J Am Chem Soc 116:1941
138. Russell VA, Ward MD (1997) J Mater Chem 7:1123
139. Horner MJ, Holman KT, Ward MD (2001) Angew Chem Int Ed 40:4045
140. Mathevet F, Masson P, Nicoud JF, Skoulios A (2002) Chem Eur J 8:2248
141. Russell VA, Evans CC, Li WJ, Ward MD (1997) Science 276:575
142. Custelcean R, Ward MD (2002) Angew Chem Int Ed 41:1724
143. Holman KT, Martin SM, Parker DP, Ward MD (2001) J Am Chem Soc 123:4421
144. Swift JA, Pivovar AM, Reynolds AM, Ward MD (1998) J Am Chem Soc 120:5887

145. Evans CC, Sukarto L, Ward MD (1999) J Am Chem Soc 121:320
146. Holman KT, Ward MD (2000) Angew Chem Int Ed 39:1653
147. Swift JA, Reynolds AM, Ward MD (1998) Chem Mater 10:4159
148. Pivovar AM, Holman KT, Ward MD (2001) Chem Mater 13:3018
149. Holman KT, Pivovar AM, Ward MD (2001) Science 294:1907
150. Mak TCW, Xue F (2000) J Am Chem Soc 122:9860
151. Félix O, Hosseini MW, De Cian A, Fischer J (1997) Tetrahedron Lett 38:1933
152. Félix O, Hosseini MW, De Cian A, Fischer J (1997) Tetrahedron Lett 38:1755
153. Félix O, Hosseini MW, De Cian A, Fischer J (1997) Angew Chem Int Ed Engl 36:102
154. Félix O, Hosseini MW, De Cian A, Fischer J (2000) Chem Commun 281
155. Hosseini MW, Brand G, Schaeffer P, Ruppert R, De Cian A, Fischer J (1996) Tetrahedron Lett 37:1405
156. Ferlay S, Félix O, Hosseini MW, Planeix J-M, Kyritsakas N (2002) Chem Commun 702
157. Burrows AD, Harrington RW, Mahon MF, Price CE (2000) J Chem Soc, Dalton Trans 3845
158. Li Q, Mak TCW (1997) Acta Crystallogr Sect B 53:252
159. Burrows AD, Mingos DMP, White AJP, Williams DJ (1996) Chem Commun 97
160. Burrows AD, Menzer S, Mingos DMP, White AJP, Williams DJ (1997) J Chem Soc Dalton Trans 4237
161. Allen MT, Burrows AD, Mahon MF (1999) J Chem Soc Dalton Trans 215
162. Burrows AD, Harrington RW, Mahon MF (2000) Cryst Eng Comm 2:66
163. Burrows AD, Harrington RW, Mahon MF, Teat SJ (2003) Eur J Inorg Chem
164. Suksangpanya U, Blake AJ, Hubberstey P, Wilson C (2002) Cryst Eng Comm 4:638
165. Kepert CJ, Hesek D, Beer PD, Rosseinsky MJ (1998) Angew Chem Int Ed 37:3158
166. Ranganathan A, Pedireddi VR, Rao CNR (1999) J Am Chem Soc 121:1752
167. Lehn J-M, Mascal M, DeCian A, Fischer J (1990) J Chem Soc, Chem Commun 479
168. Zerkowski JA, Seto CT, Wierda DA, Whitesides GM (1990) J Am Chem Soc 112:9025
169. Zerkowski JA, MacDonald JC, Seto CT, Wierda DA, Whitesides GM (1994) J Am Chem Soc 116:2382
170. Zerkowski JA, Whitesides GM (1994) J Am Chem Soc 116:4298
171. Zerkowski JA, Seto CT, Whitesides GM (1992) J Am Chem Soc 114:5473
172. Zerkowski JA, Mathias JP, Whitesides GM (1994) J Am Chem Soc 116:4305
173. Zerkowski JA, MacDonald JC, Whitesides GM (1994) Chem Mater 6:1250
174. Steinbeck M, Ringsdorf H (1996) Chem Commun 1193
175. Bohanon TM, Denzinger S, Fink R, Paulus W, Ringsdorf H, Weck M (1995) Angew Chem Int Ed Engl 34:58
176. Bohanon TM, Caruso P-L, Denzinger S, Fink R, Möbius D, Paulus W, Preece JA, Ringsdorf H, Schollmeyer D (1999) Langmuir 15:174
177. Champ S, Dickinson JA, Fallon PS, Heywood BR, Mascal M (2000) Angew Chem Int Ed 39:2716
178. Kawasaki T, Tokuhiro M, Kimizuka N, Kunitake T (2001) J Am Chem Soc 123:6792
179. Fouquey C, Lehn J-M, Levelut A-M (1990) Adv Mater 2:254
180. Kimizuka N, Kawasaki T, Hirata K, Kunitake T (1995) J Am Chem Soc 117:6360
181. Kimizuka N, Fujikawa S, Kuwahara H, Kunitake T, Marsh A, Lehn J-M (1995) J Chem Soc Chem Commun 2103
182. Choi IS, Li X, Simanek EE, Akaba R, Whitesides GM (1999) Chem Mater 11:684
183. Gulik-Krzywicki T, Fouquey C, Lehn J-M (1993) Proc Nat Acad Sci USA 90:163
184. Beijer FH, Sijbesma RP, Vekemans JAJM, Meijer EW, Kooijman H, Spek AL (1996) J Org Chem 61:9636
185. Lange RFM, Beijer FH, Sijbesma RP, Hooft RWW, Kooijman H, Spek AL, Kroon J, Meijer EW (1997) Angew Chem Int Ed Engl 36:969
186. Bishop MM, Lindoy LF, Skelton BW, White AH (2002) J Chem Soc Dalton Trans 377
187. Houlton A, Mingos DMP, Williams DJ (1994) J Chem Soc Chem Commun 503
188. Houlton A, Mingos DMP, Williams DJ (1994) Transition Met Chem 19:653
189. Bernhardt PV (1999) Inorg Chem 38:3481
190. Burrows AD, Mingos DMP, White AJP, Williams DJ (1996) J Chem Soc, Dalton Trans 149

191. Chowdhry MM, Burrows AD, Mingos DMP, White AJP, Williams DJ (1995) J Chem Soc Chem Commun 1521
192. James SL, Mingos DMP, Xu XL, White AJP, Williams DJ (1998) J Chem Soc Dalton Trans 1335
193. Chowdhry MM, Mingos DMP, White AJP, Williams DJ (1996) Chem Commun 899
194. Lehn J-M, Mascal M, DeCian A, Fischer J (1992) J Chem Soc Perkin Trans 2:461
195. Mascal M, Hext NM, Warmuth R, Moore MH, Turkenburg JP (1996) Angew Chem Int Ed Engl 35:2204
196. Falvello LR, Garde R, Tomás M (2002) Inorg Chem 41:4599
197. Mascal M, Fallon PS, Batsanov AS, Heywood BR, Champ S, Colclough M (1995) J Chem Soc Chem Commun 805
198. Mascal M, Hansen J, Fallon PS, Blake AJ, Heywood BR, Moore MH, Turkenburg JP (1999) Chem Eur J 5:381
199. Fan E, Yang J, Geib SJ, Stoner TC, Hopkins MD, Hamilton AD (1995) J Chem Soc Chem Commun 1251
200. Hassaballa H, Steed JW, Junk PC (1998) Chem Commun 577
201. Schmuck C, Wienand W (2001) Angew Chem Int Ed 40:4363
202. Beijer FH, Kooijman H, Spek AL, Sijbesma RP, Meijer EW (1998) Angew Chem Int Ed 37:75
203. Hirschberg JHKK, Brunsveld L, Ramzi A, Vekemans JAJM, Sijbesma RP, Meijer EW (2000) Nature 407:167
204. Sharma CVK, Clearfield A, Cabeza A, Aranda MAG, Bruque S (2001) J Am Chem Soc 123:2885
205. Sánchez L, Rispens MT, Hummelen JC (2002) Angew Chem Int Ed 41:838
206. Hirschberg JHKK, Beijer FH, van Aert HA, Magusim PCMM, Sijbesma RP, Meijer EW (1999) Macromolecules 32:2696
207. Berl V, Schmutz M, Krische MJ, Khoury RG, Lehn J-M (2002) Chem Eur J 8:1227
208. Plaut DJ, Holman KT, Pivovar AM, Ward MD (2000) J Phys Org Chem 13:858

Structure and Bonding, Vol. 108 (2004): 97–168
DOI 10.1007/b14138

Molecular Containers: Design Approaches and Applications

David R. Turner[1] · Aurelia Pastor[2] · Mateo Alajarin[2] · Jonathan W. Steed[1]

[1] Department of Chemistry, University Science Laboratories, University of Durham, South Road, Durham, DH1 3LE, United Kingdom
E-mail: jon.steed@kcl.ac.uk

[2] Departamento de Química Organica, Facultad de Química, Universidad de Murcia, Campus de Espinardo, Murcia-30.100, Spain

Abstract The design and synthesis of molecular containers is playing an increasing role in the selective removal and detection of species within solution. The cavities offered by such species provide the possibility of three-dimensional molecular recognition and therefore highly selective host species. Many varied approaches towards the design of container compounds have been adopted, ranging from rigid, covalently formed carcerands to self-assembling dimers and oligomers. This chapter explores the wide range of approaches possible; covalently formed containers, cages assembled around metal centres and those which self-assemble via non-covalent interactions. The main uses of such systems, for stabilising reactive species and promoting reactions within the protective environment of cavities, are also highlighted.

Keywords Self-assembly · Hydrogen-bonding · Host-guest · Encapsulation · Capsule

1 Introduction 98

2 Covalently Formed Capsules 99

2.1 Cryptands and Cyclophanes 99
2.2 Carcerands 100
2.3 Hemicarcerands and Cryptophanes 104

3 Metal Directed Self-Assembling Cages 109

3.1 Metal Directed Synthesis 109
3.2 Assembly via Scaffolding Ligands 110
3.3 Assembly via Panelling Ligands 117

4 Non-Covalent Assemblies 133

4.1 Multi-Component Assemblies 133
4.2 Self-Complimentary Capsules 140
4.3 Glycoluril Systems 146
4.4 Urea Containing Capsules 153
4.5 Unimolecular Capsules 162

5 Conclusions and Outlook 164

6 References 164

List of Abbreviations

en Ethylenediamine
acac Acetyl acetonate
ESI Electrospray ionisation
FAB Fast atom bombardment
NFP *N*-Formylpiperidine
NMP *N*-Methylpyrrolidinone
NOE Nuclear Overhauser effect

1 Introduction

Molecular containers, completely enclosed hollow species capable of holding one or more guest species inside, are becoming of increasing interest for molecular recognition [1–4]. The ability to recognise and detect molecules accurately has many potential practical applications for the sensing [5] and sequestration of species present within solutions. Containers, or capsules, provide ideal structures for recognition as they have the potential to provide discrimination in terms of size, shape and functionalities within a 3D space. The interior of a cavity can also stabilise reactive species by isolating guests from the bulk environment and can catalyse reactions effectively due to guest discrimination.

The approaches taken to the design and synthesis of container molecules are numerous and varied (Fig. 1). Containers can be made as single, large covalently joined molecules [6, 7]. More common approaches in recent years have been

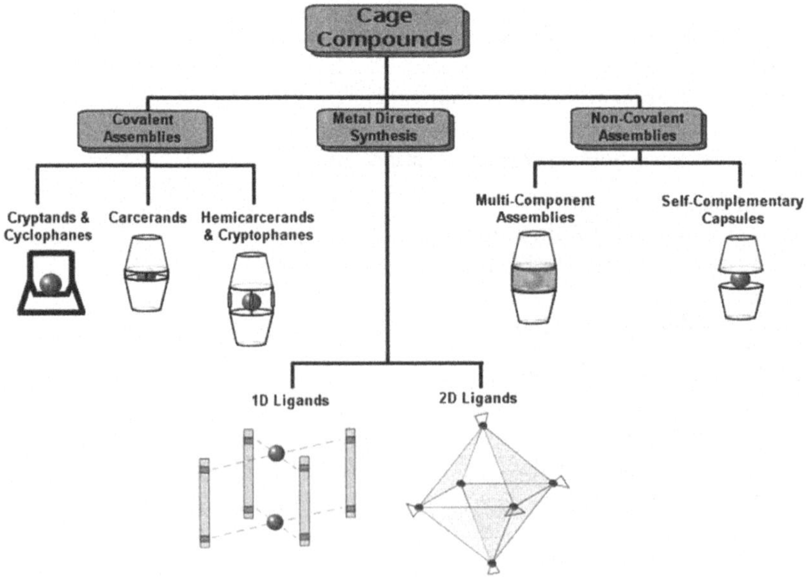

Fig. 1 Methods for the assembly of molecular containers

focused around the self-assembling of several species to form cage-type compounds [8,9]. The use of metal directed assembling techniques provides a great deal of versatility in the construction of complex geometries [10–20]. The formation of capsules can also be achieved via non-covalent assemblies of molecules which can assemble in solution around the guests [21–25].

2 Covalently Formed Capsules

2.1 Cryptands and Cyclophanes

The covalent assembly of guest-encapsulating host species has long been an area of research interest. The first examples of host species binding their guests within a three-dimensional array of interactions were the class of compounds known as cryptands [26,27]. The cryptands were designed as hosts for alkali metal cations and are based on macrobicyclic-polyethers (Fig. 2). Typically, cryptands are synthesised by the addition of a diacyl-chloride to an azacrownether (Scheme 1).

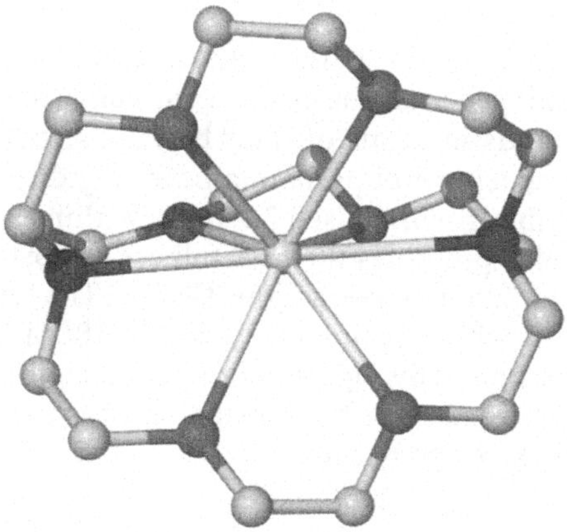

Fig. 2 Schematic representation of [2,2,2]-cryptand binding K^+ in a six-fold array of interactions with ether oxygen atoms and also via nitrogen interactions

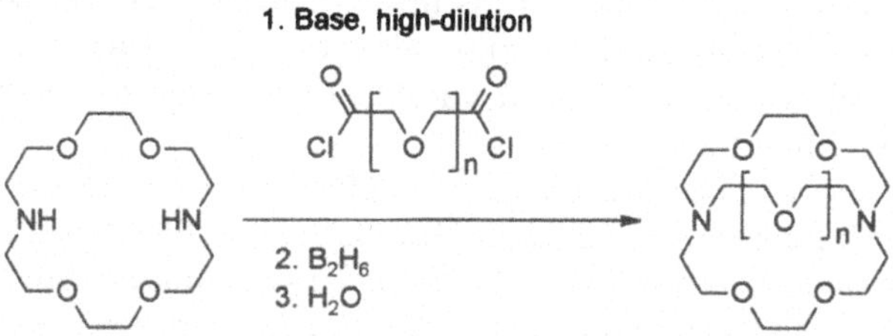

Scheme 1 The synthesis of simple cryptands via the high-dilution addition of acyl chlorides to azacrownethers

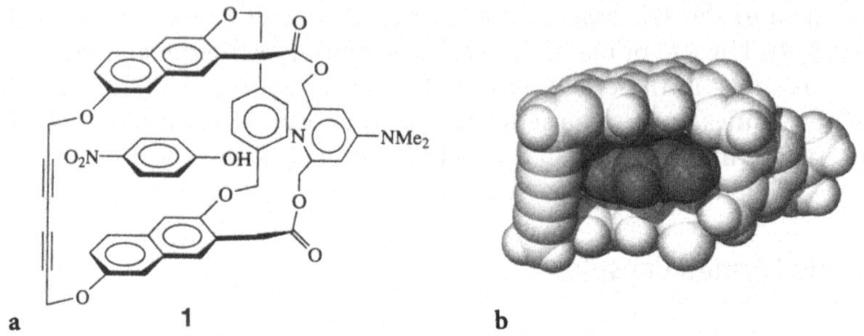

Fig. 3 a The *p*-nitrophenol guest resides within the cavity of the cyclophane **1** via π-π stacking. b The crystal structure showing the inclusion of the guest within the cavity

The cryptands are very adaptable simply by changing the size and substituents of the macrocycles from which they are composed. The complexation of cationic guests within the cavities offered by cryptands is significantly stronger than comparable hosts (i.e. crown ethers and lariat ethers) due to the preorganisation of the 3D binding site. Such hosts are, however, limited to a small size, since with increasing chain length comes a decrease in conformational rigidity and loss of the preorganised cavity shape.

A related class of compounds are the larger and more rigid cyclophane hosts (cyclic species containing at least one bridged aromatic ring). Many of these have a similarity to cryptands in terms of their bicyclic structure [28]. The use of alkyne and aromatic spacing groups maintains the rigidity of the hosts and presents an easily accessible, preorganised 3D space within their confines. The use of aromatic spacers also promotes their usefulness as hosts for aromatic guests by virtue of the substantial π-π interactions (Fig. 3). The binding constant for *p*-nitrophenol as a guest within cyclophane **1** is $9.6\times10^4\ M^{-1}$ (in CH_2Cl_2 at room temperature), for example. This figure represents a significant interaction between two neutral species, despite the substantial gaps in the sides of the cavity through which guest exchange occurs.

2.2 Carcerands

Some of the earliest work to be carried out on cage compounds was by Donald Cram and co-workers [7]. Much of his initial work was concerned with bowl-shaped species containing deep cavities, called cavitands, such as **2** (Fig. 4). Such hosts are capable of strong neutral guest complexation due to the deep hydrophobic cavity that they possess.

Later in his career Cram designed and synthesised a new class of compounds that he termed 'carcerands'. These compounds are completely enclosed, roughly spherical cages with an interior cavity that is large enough to bind simple guest species [29]. The term carcerand comes from the fact that once the guest species are incarcerated within the host, they cannot escape. Carcerands that are occupied with guest species are termed 'carceplexes'. The synthesis of Cram's

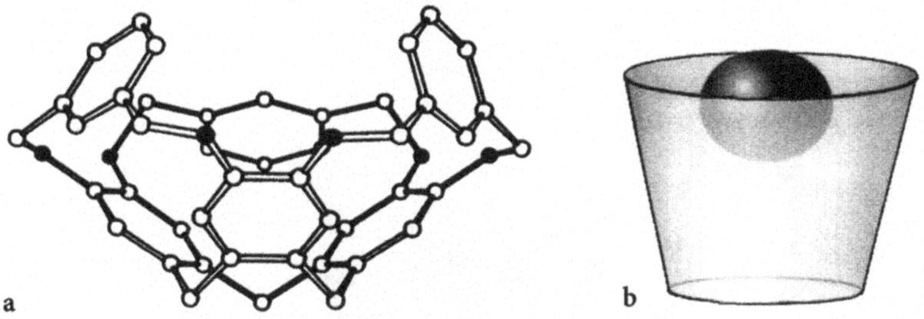

Fig. 4 **a** A crystal structure of the cavitand **2**, which can be likened to a deep bowl **b** in which a guest can reside [7]. Reproduced by permission of The Royal Society of Chemistry

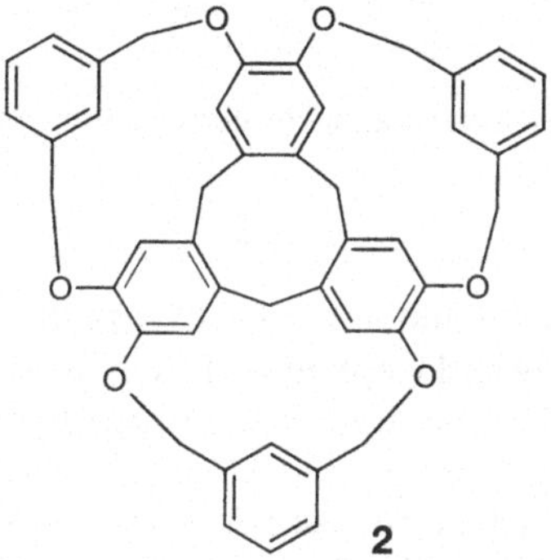

2

carcerands is via the joining of two hemi-spherical components in a covalent manner to give a sulphur bridged equatorial seam (Scheme 2). Once the cage is formed the gaps left in the structure are too small for any encapsulated guest larger than water to escape (Fig. 5).

Carcerand **3** is an extremely insoluble compound that led to a challenging analysis for Cram and his co-workers. Work had to be carried out solely by

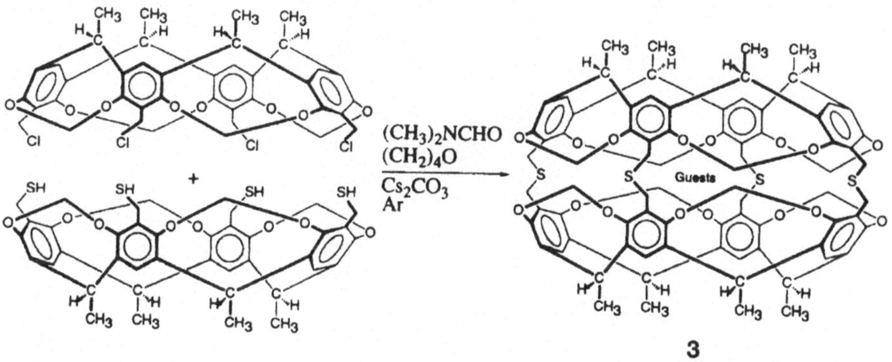

3

Scheme 2 The synthesis of a carcerand, **3**, from the joining of two differently substituted, deep cavity species via thioether bridges [7], reproduced by permission of The Royal Society of Chemistry

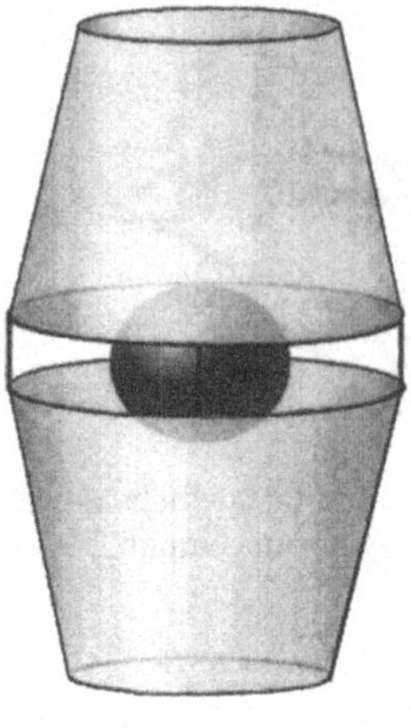

Fig. 5 Cartoon representation of a carcerand. The guest cannot escape through any gaps in the cage structure

mass spectrometry and elemental analyses. Eventually it transpired that as the carcerands had formed they had trapped within them any species that happened to be present in the reaction medium at the time of cavity closure. Later work dubbed this the guest determining step (GDS) [6]. Reactants, solvent molecules and even the argon under which the reaction was carried out were found to be present in various amounts within the carceplex samples obtained. Cram was able to provide mechanistic evidence that the location of some of the species immediately prior to the closure of the carcerand led to increased proportions of certain included atoms (Fig. 6) [30]. Once these guests were inside the carcerands they could not leave again until the complex was subjected to a chem-

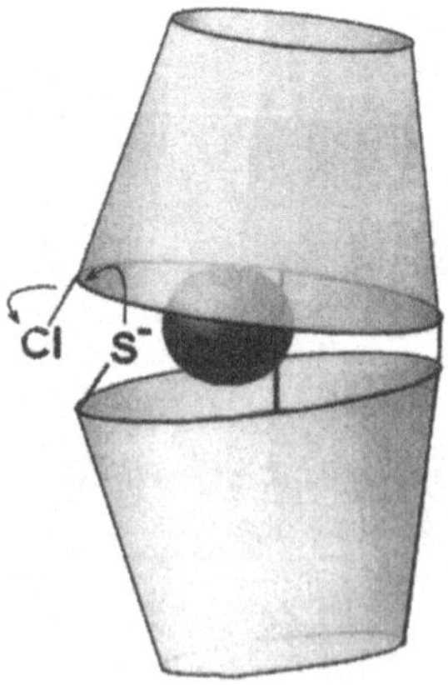

Fig. 6 The mechanism by which the $Cs^+S^-_{cavitand}$ ion pair on the lower hemisphere reacts with the CH_2Cl group of the upper hemisphere in an S_N2 manner during the cage closure leads to a large amount of Cs^+ being present within the carceplex in the final samples

ical method of breaking some of the cage forming bonds. The only species that was able to enter the carceplex after it had formed was water if it was subjected to sufficient pressure to overcome the complexation activation energy. There was no evidence that it can escape again. This is understandable as the reverse process would require pressure to be exerted from the inside.

In order to overcome some of the problems with the handling and characterisation of the compounds, long *n*-alkyl chains were placed around the exterior of the cage to increase the solubility of the carceplexes. Using these more soluble compounds, Cram was able to discover that the carcerands were selectively trapping guests of adequate size and no carceplex was formed when the reaction was run in NFP, a molecule too big for the interior of the carcerand. This indicates a process of guest-controlled assembly, whereby carceplexes do not form unless a suitable template is present, a technique often employed when synthesising hosts or assemblies under kinetically controlled reaction conditions [31]. Further studies have shown that the template effect varies one million-fold, with pyrazine as the best template and NMP as the poorest measurable template.

Only one example of a guest escaping from a carceplex-type system has been documented. When a carceplex containing two molecules of acetonitrile was heated at 110 °C for 72 h, one of the molecules of acetonitrile was observed to escape. The ejection process was followed at different temperatures using ^{1}H-NMR and it was found that the activation energy for this process was 20 kcal mol^{-1}. This was attributed to a 'billiard-ball effect', whereby the two incarcerated molecules collide to provide one very high energy species. If this occurs when the molecules are correctly aligned then the high energy species may escape. This escape phenomenon was not observed when only one molecule was incarcerated [32].

Carcerands have also been prepared with OCH_2O bridging units instead of the CH_2SCH_2 thioether bridges. The oxygen bridged carcerands have a slightly

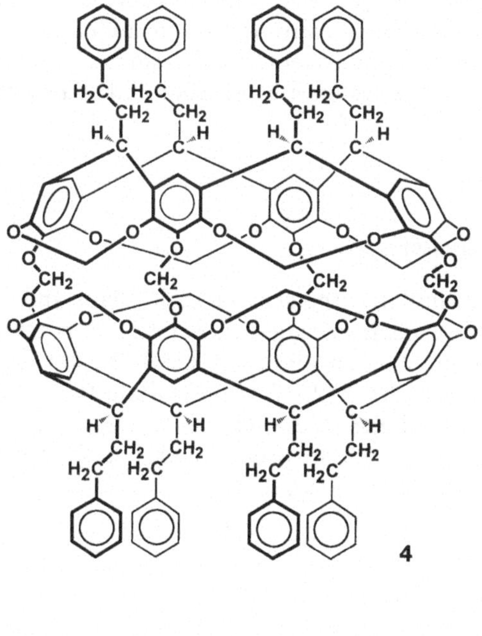

4

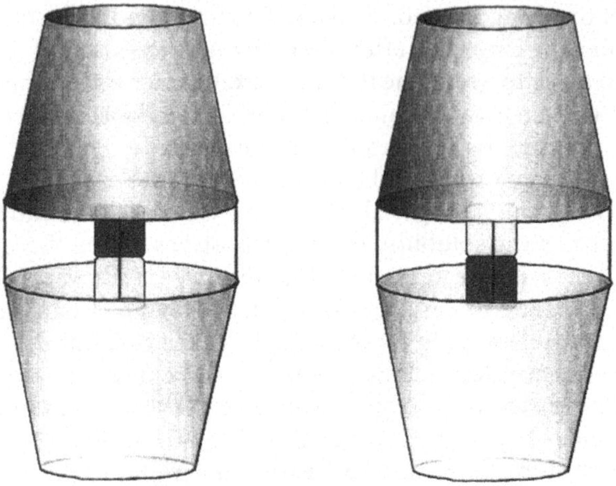

Fig. 7 Schematic representation of a pair of carceroisomers

smaller volume than the thioether species and are observed to incarcerate small solvent molecules. Evidence of guest inclusion was also observed in the solid state for **4** with encapsulated $(CH_3)_2NCOCH_3$ [33].

Reinhoudt has reported the first carcerand built from two different hemispheres [34]. These carceplexes exhibit a novel type of stereoisomerism as a result of different orientations of the guest molecule inside the cavity (Fig. 7). They tentatively proposed the name of 'carceroisomerism', referring to the hindered rotation of the molecule inside the cavity of the carcerand. Although carcerands are capable of binding guests with effectively infinite strength (as there is no escape), they bind unselectively and there is no possible exchange of guest species. Encapsulation can also only occur under the reaction conditions of the original synthesis. Such systems are therefore of limited use in the main applications of host-guest systems; as sensing and sequestering agents or as chemical catalysts.

2.3 Hemicarcerands and Cryptophanes

Hemicarcerands are similar to the carcerands in shape and structure with one major difference – the capacity for guest exchange. This is achieved in one of two ways; either via a portal made by omitting one of the four bridging groups between the hemispheres (Fig. 8) [35] or by making the bridging groups long enough to provide large holes in the side of the cage (Fig. 9) [36]. As with the carcerands, the hemicarcerands are produced via a templated reaction and there is no empty hemicarcerand formed.

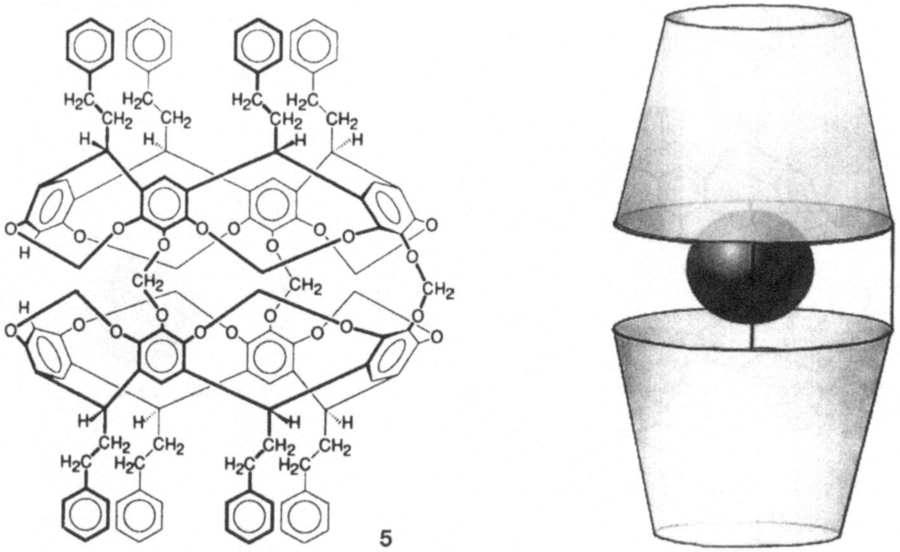

Fig. 8 A hemicarcerand with only three connecting units and a single portal [7], reproduced by permission of The Royal Society of Chemistry, and a cartoon representation of the hemicarceplex

The guest exchange kinetics of the hemicarceplexes with a single portal were monitored by ^{1}H-NMR spectroscopy. The temperatures at which decomplexation was found to occur were related to the size and shape fit of the guest compared to the portal through which the exchange was taking place. Small molecules, such as diatomic gases and water, can enter and leave relatively easy, whereas larger solvent molecules such as chloroform are sterically prevented from doing so. At 140 °C, solvent molecules displayed first order behaviour with long half-lives, indicating a slow exchange process. With host **5**, half-lives of 14 h and 34 h were observed for $(CH_3)_2NCHO$ and $(CH_3)_2NCOCH_3$ respectively, using 1,2,4-trichlorobenzene as a solvent (this eliminates the possibility of solvent inclusion on steric grounds). However, modelling of these processes showed that neither of these guest species can enter or leave the host without bonds of the host being broken. Exchange at ambient temperatures, therefore, is unlikely to occur with such guests. Smaller solvent molecules, such as acetonitrile, have been observed to show guest exchange behaviour at ambient temperatures. The half-life of the hemicarceplex of **5** and acetonitrile in a 1:1 ratio is 13 h at 22 °C in dichloromethane. Larger guests, such as benzene, can be incorporated but only at very high temperatures. The selectivity of **5** in terms of size exclusion is good but it is not possible to discriminate between molecules of a similar size very easily. In a study using O_2, N_2 and H_2O the ^{1}H-NMR spectra showed a slow exchange between all three species, as well as signals representing the free host. Observation of the free host is possi-

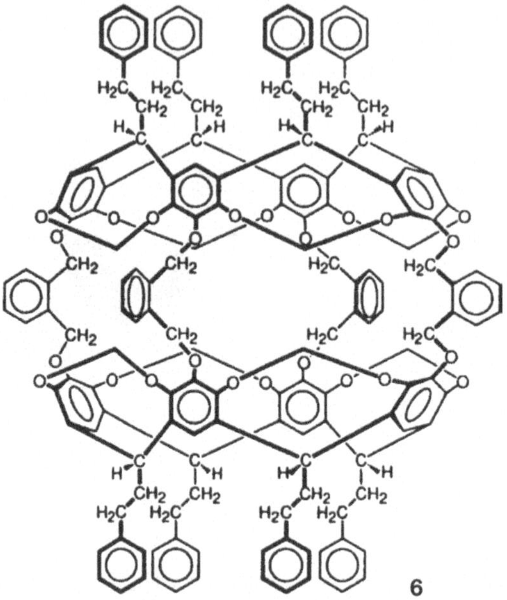

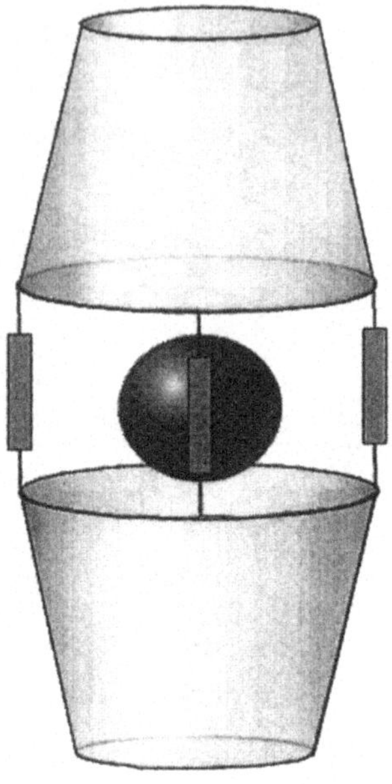

Fig. 9 A hemicarcerand with extended bridging units, offering four portals [7], reproduced by permission of The Royal Society of Chemistry, and a cartoon representation of the hemicarceplex

ble because with one lone portal it is necessary for one guest to exit the cavity before the next can enter, as opposed to a concerted process.

The second class of hemicarcerands, those containing four large entries offered by extending the bridging groups, displays different guest exchange properties. Guests as large as *p*-xylene are able to gain access to the cavity of **6** and show slow exchange properties. Discrimination due to the geometry of the guest is also evident, as *o*- and *m*-xylene were not complexed. As with the single-portal hemicarcerands, the exchange of guests occurs via a two-step process. Half-lives for hemicarceplexes of **6** ranged from 38 min for acetonitrile to 6.5 h for ethyl acetate in 1,2-dichloroethane.

The variety of hemicarcerands that have now been prepared and studied runs into the hundreds [6, 37]. Differing functionalities can be placed inside the cavity and the size of the cavity itself is easily adjusted by changing the bridging groups between the two hemispheres. What the hemicarcerands fail to offer, though, is the selectivity that is required to make hosts for practical applications involving guest recognition. The large gaps possessed by hemicarcerands do not provide sufficient discrimination between potential guest species.

Hemicarcerands do, however, show interesting behaviour in terms of the environment within the cavity. Highly reactive species can be stabilised within

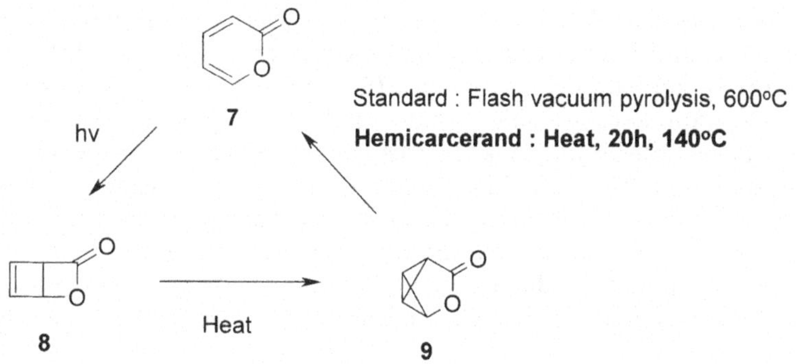

Scheme 3 The interconversions of lactones, **8** and **9**, and α-pyrone **7** is accomplished under milder conditions when carried out within a hemicarcerand than when carried out in solution

the capsule and the interior space can act as a catalytic chamber, to promote reactions which cannot occur under normal conditions. An example of this behaviour is the pyrone and lactone interconversions shown in Scheme 3. α-Pyrone **7** is converted photochemically to give **8**, which is then heated to form **9**. When this reaction sequence is conducted within the confines of a hemicarcerand the final conversion back to **7** occurs under much milder conditions due to the unique environment that the cage interior possesses [38]. Other remarkable phenomena displayed by hemicarcerands include the stabilisation of the highly unstable cyclobutadiene (derived from the photolysis of **8**) [38], stabilisation of *o*-benzyne [39] and an 'innermolecular' Diels-Alder reaction [40]. Through shell reactions occurring between guests in the inner phase of the hemicarcerand and a reagent dissolved in the outer-phase solvent have also been described, such as oxidation reactions to give unstable benzoquinones, the reduction of $PhNO_2$ to give PhNHOH rather than $PhNH_2$ or bimolecular S_N2 O-methylation of phenols with outer-phase reactants [41, 42].

Carceroisomerism has also been observed in hemicarceplexes. Paek and co-workers have measured isomerisation energy barriers of carceroisomers in non-centrosymmetric C_{4v} hemicarceplexes, the largest of which was found to be 15.4 Kcal mol^{-1} for the rotation of NMP inside the cavity [43]. It has also been claimed that the inside of carcerands and hemicarcerands can be considered as a new phase of matter. This suggestion implies effects beyond mere spatial confinement and chemical isolation, for example, a marked change in the physical bulk properties, such as the polarity or polarisability of the host cavity. Nau has obtained evidence that biacetyl included within the cavity of a hemicarcerand may experience an unusual polarisability even higher than that of diiodomethane by using biacetyl as a solvatochromatic probe for the polarisability of the environment [44].

Cryptophanes are related to hemicarcerands. They are composed of smaller, shallower cavitand bowls, separated by long spacer groups. Two different conformers of cryptophanes can exist, *anti* **10** and *syn* **11** (X=alkyl or aryl group) [45, 46]. Cryptophanes are able to bind methane and its halogenated derivatives well. Reversible binding has also been observed with Xe by means of ^{129}Xe-NMR studies [47]. The exchange is slow due to the restricted movement of the guest through the portals of the host. The hosts are often water-soluble when hydrophilic groups replace the methoxy ones and guest binding is enhanced in such a medium due to the hydrophobic interior of the cavity. Cryptophane **12**, for example, displays a binding constant of 7700 M^{-1} for $CHCl_3$ in water.

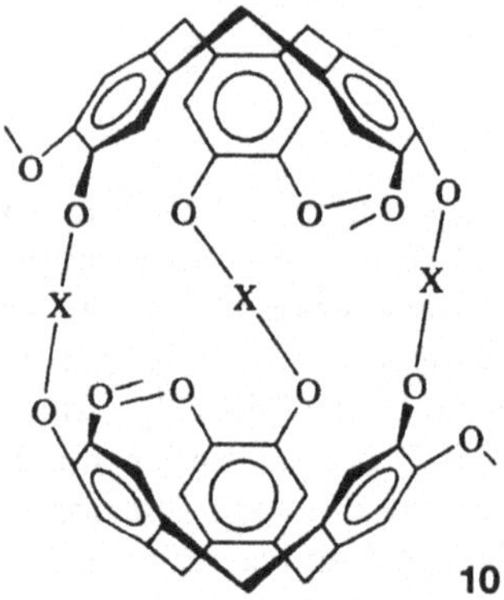

10

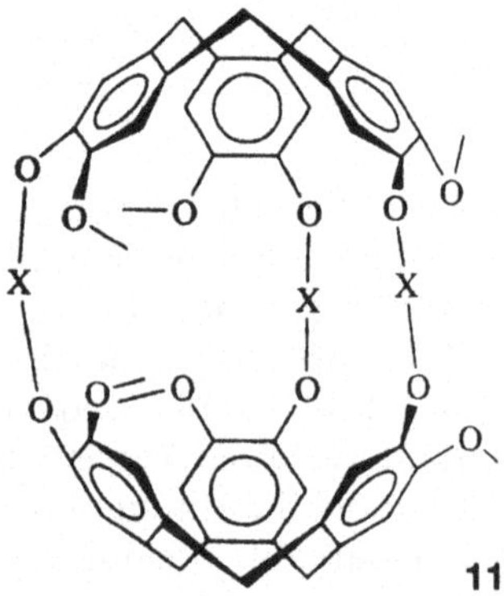

11

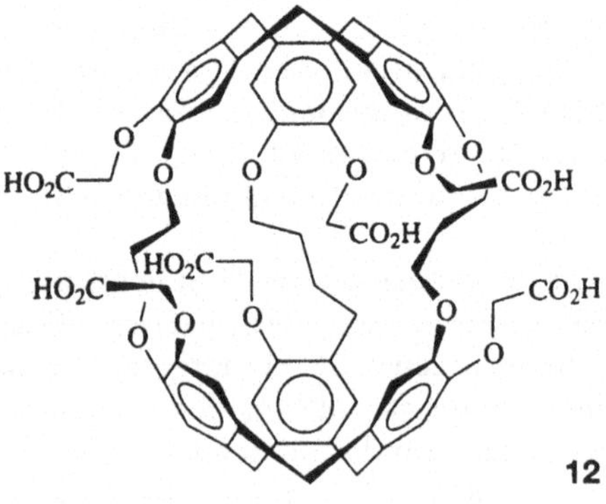

12

3
Metal Directed Self-Assembling Cages

3.1
Metal Directed Synthesis

Self-assembly is the spontaneous coming together of several chemical entities to form a larger aggregate under thermodynamic control [8]. With a careful and informed choice of starting materials the resulting compounds can often be predicted and therefore design strategies can be formulated. Self-assembly as a synthetic method is most commonly applied to coordination compounds and is frequently termed 'metal-directed synthesis'.

Metal centres, with their strict coordination geometries, provide ideal building blocks for producing complex 3D shapes. When coupled with rigid spacing units, either 1D (to produce a scaffold style structure, Fig. 10) or 2D (molecular panelling, Fig. 11), a wide variety of geometrical shapes are accessible [12, 17, 20]. The coordination topology of the metal centre is easily controllable by alteration of the lability of the ligands attached to the starting materials. The organic ligand can also be controlled by the positioning of the interaction sites.

The thermodynamically controlled synthesis gives rise to another advantage of metal-directed assembly: the capability to correct mistakes within the assembly until the final product is formed. The reversibility of M-L bond formation means that large assemblies and the individual components are in equilibrium until a stable product is formed. Many small assemblies form in the reaction mixture, some of which continue to grow towards the final product while others fall apart and their components are recycled.

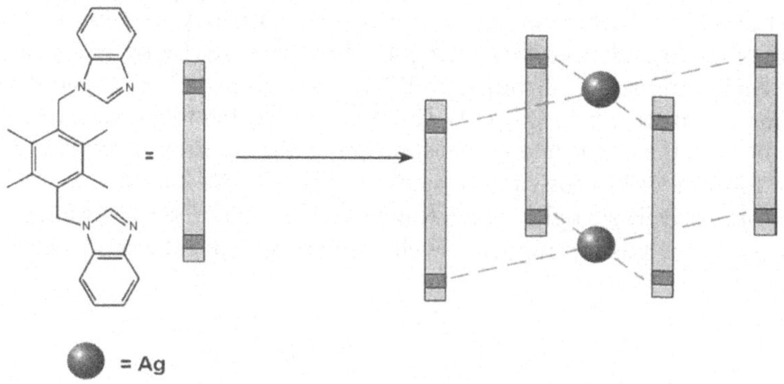

Fig. 10 The use of 'scaffolding' ligands combined with metals to create a 3D cage structure, after Su et al. [54]

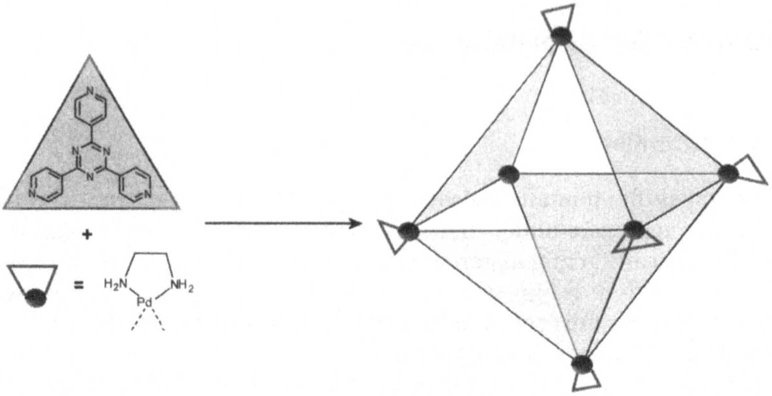

Fig. 11 'Molecular panelling', the construction of 3D shapes via the use of polygonal ligands and semi-protected metals, after Fujita et al. [61]

3.2 Assembly via Scaffolding Ligands

Roughly linear or thread-like ligands can be used to form the edges of simple polyhedra when combined with the appropriate metal complexes. Although the vast majority of ligands used are not strictly one dimensional, for the purposes of this work the term '1D ligands' can be taken to mean ligands that only bridge between two points and can therefore be simply thought of as topologically equivalent to straight lines. Ligands of this type assemble into cages by forming the edges of polyhedra to produce a scaffold-type structure.

In some instances, this approximation does not provide a very accurate model, such as those cases in which the resulting compounds display a helical structure with bent ligands, such as the cage $[Pd_2(1,4\text{-bis}(3\text{-pyridyloxy})\text{benzene})_4]$ **13** [48]. The nature of the ligand leads to the two palladium (II) ions being staggered with respect to each other by 45° through the Pd-Pd axis, giving a helical geometry, as seen in the crystal structure (Fig. 12). Complex **13** has been shown to enclose a PF_6^- anion both in the solid state and in solution. In the solution phase the anionic guest is in motion within the cage as evidenced by ^{19}F-NMR data.

The use of relatively straight connecting ligands enables the formation of standard polyhedra, such as tetrahedra. A well studied example of a tetrahedral cage

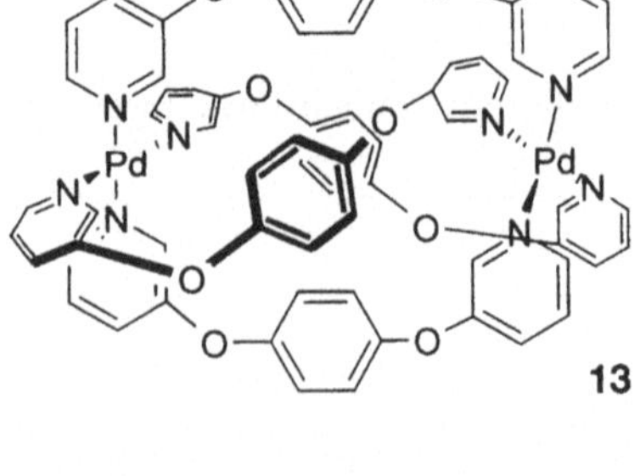

13

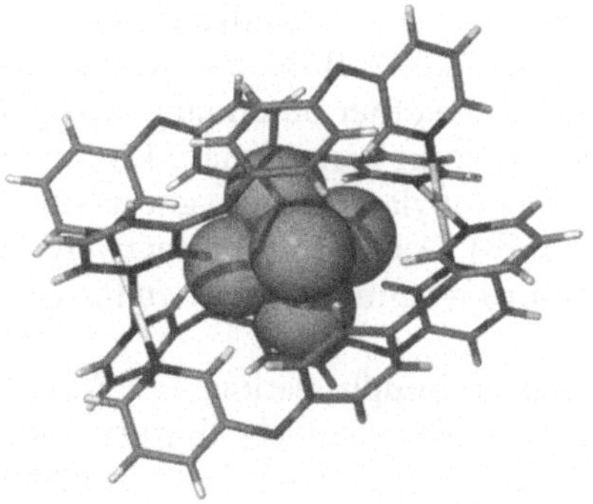

Fig. 12 A helical Pd based cage and its solid state encapsulation of PF_6^-, (non-encapsulated anions are not shown for clarity)

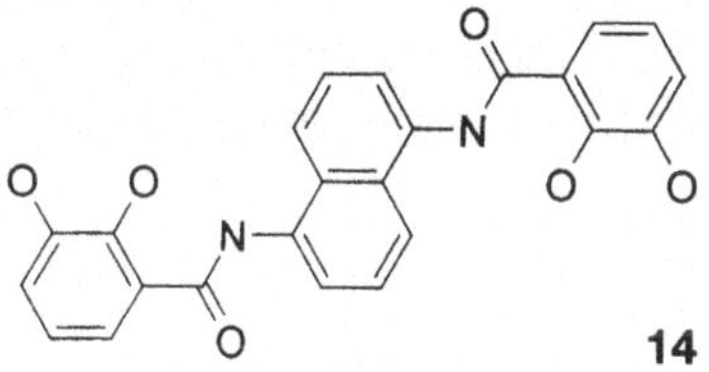

14

is that developed by Raymond et al. [49, 50]. The bis-bidentate ligand **14** and four gallium or iron atoms self-assemble to form an M_4L_6 capsule with a large negative charge making it ideal for encapsulating positively charged guests in both solution and the solid state (Fig. 13). The flexibility of the ligand allows for slight changes in the shape of the cage to accommodate a range of guest species. The water soluble cage is capable of encapsulating a range of tetraalkyl ammonium cations (Me_4N^+, Et_4N^+, Pr_4N^+). When tetramethyl ammonium is used as a guest the exchange is fast on an NMR timescale, whereas the larger guest species dis-

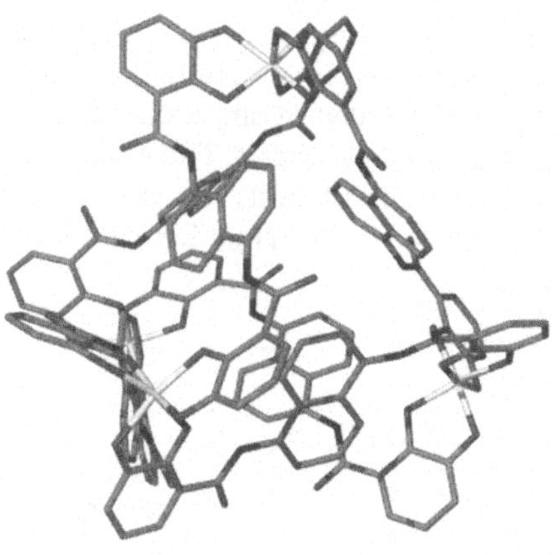

Fig. 13 Crystal structure of an $[Fe_4(\mathbf{14})_6]^{12-}$ tetrahedral cage, counter-ions and hydrogen atoms not shown for clarity

play slow exchange. The cage is able to discriminate between different NR_4^+ guests in the presence of a mixture. If the methylammonium complex is placed in the presence of either of the other potential guest species (Et_4N^+, Pr_4N^+), the methyl substituted guest is completely replaced in under one minute. Binding constants of the ethyl- and propylammonium guests (in D_2O, relative to the 12 K^+ complex) are 1.96×10^4 and 1.11×10^2 M^{-1}, respectively. Tetraethylammonium is observed to replace tetrapropylammonium when the two cations are both in the presence of the host.

As well as encapsulating simple cations, an $M_4(\mathbf{14})_6$ cage has been observed to bind highly unstable positively charged species [51]. For example, $[Me_2C(OH)PEt_3]^+$, **15**, had only previously been isolated under anhydrous conditions as it rapidly decomposes in the presence of water (Scheme 4). On the addition of PEt_3 to $[Ga_4(\mathbf{14})_6]^{4+}$, new signals assigned to **15** were observed in the NMR spectra. This is accounted for by the intracavity formation of **15** when protonated PEt_3 diffuses into the cavity and reacts with acetone that remains in the cavity after the initial synthesis. The resulting complex remains stable in D_2O for several hours. This was the first example of guest stabilisation in a supramolecular metal cage.

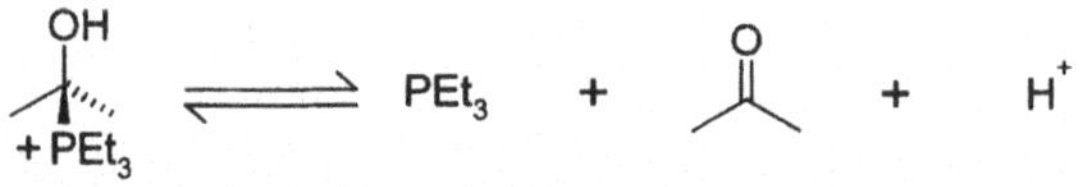

15

Scheme 4 The decomposition of $[Me_2C(OH)PEt_3]^+$ in water, which is stabilised by inclusion in $[Ga_4(\mathbf{14})_6]^{12-}$

It has also been demonstrated in similar systems that the guest species can have a profound influence on the structure of the cage itself [52]. Ligand **16** is an extended version of **14** with an anthracenyl spacer instead of naphthyl. When assembled together with $[TiO(acac)_2]$ and KOH a triple helicate M_2L_3 structure is formed, detected by ESI-MS and single crystal X-ray analysis (Fig. 14a). When $[TiO(acac)_2]$ and **16** are reacted using Me_4NOH the helicate does not form, instead the tetrahedral cage is isolated with one encapsulated cation (Fig. 14b). This is an example of guest-templated synthesis. The analogous gallium compounds were also studied, due to the greater lability of gallium. It was found that the addition of Me_4NCl to a solution of the triple helicate led to a change, over the

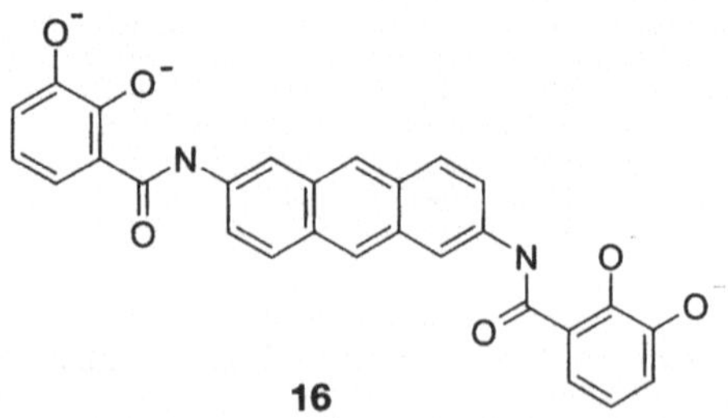

16

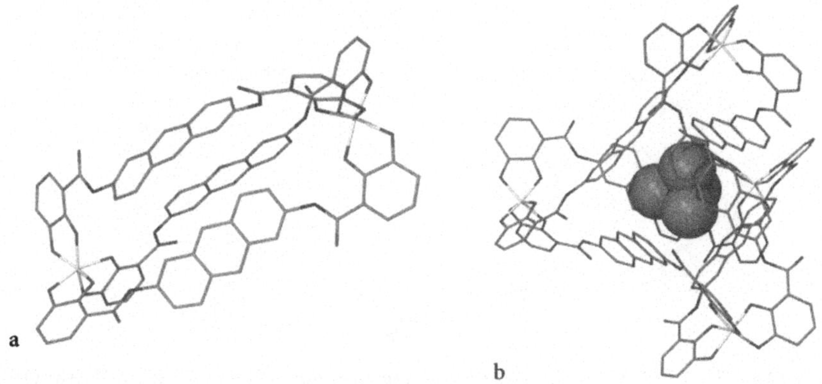

Fig. 14a, b Crystal structures of: **a** $Ti_2(\mathbf{16})_3$ (triple helical); **b** $Ti_4(\mathbf{16})_6 \cdot Me_4N^+$ (tetrahedral) complexes, non-coordinating counter-ions are not shown

course of five days, into the tetrahedral cage. This cage has also been observed to bind the K^+ complex of [12]crown-4 [53].

Anionic guest species also play an important role within the chemistry of metal-ligand cages. Trigonal pyramidal and tetrahedral cages have been synthesised from ligands **17** and **18** with $AgCF_3SO_3$ and $Cu(ClO_4)_2$ respectively, in which the encapsulated anion is itself interacting with the metal centres [54].

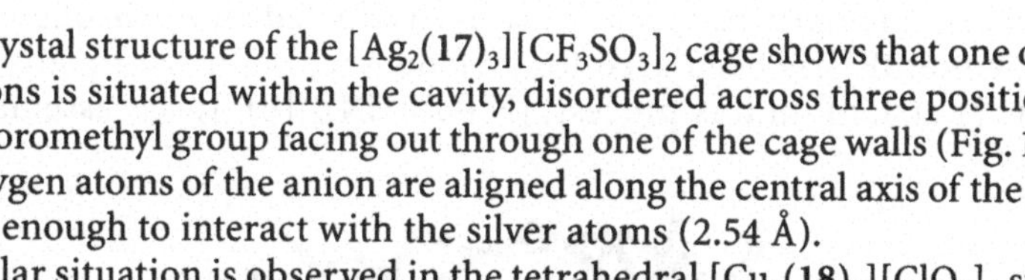

17 **18**

The crystal structure of the $[Ag_2(\mathbf{17})_3][CF_3SO_3]_2$ cage shows that one of the triflate anions is situated within the cavity, disordered across three positions, with the trifluoromethyl group facing out through one of the cage walls (Fig. 15a). Two of the oxygen atoms of the anion are aligned along the central axis of the cage and are close enough to interact with the silver atoms (2.54 Å).

A similar situation is observed in the tetrahedral $[Cu_2(\mathbf{18})_4][ClO_4]_4$ structure, where two of the perchlorate oxygen atoms are found to be interacting with the Cu atoms in the solid state (Fig. 15b). Neither of these cages have been explored in terms of their guest exchange properties. It appears, however, that their assembly is templated by the presence of the central anion.

The templation of metal-ligand assemblies by anions can occur in two ways: a Lewis acid/base interaction between the anion and the metal or via hydrogen bonding between the anion and organic ligands. One system that displays both of these interactions is the halide templated assembly between **19**, one of the tautomeric forms of amidinothiourea (Scheme 5), and NiX_2 (X=Cl, Br) [55]. In the

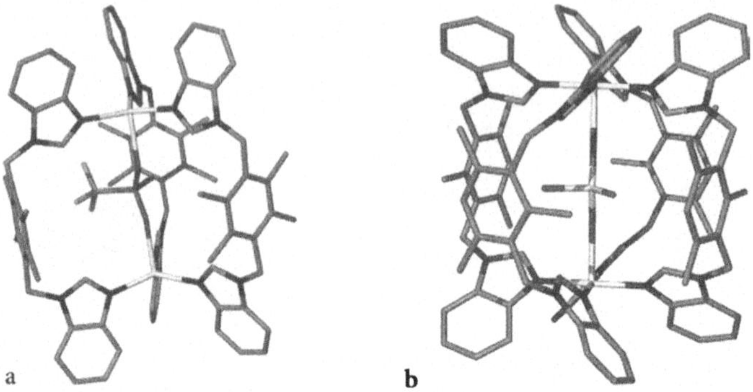

a b

Fig. 15 a $[Ag_2(17)_3][CF_3SO_3]_2$ with a triflate anion interacting with the Ag centres. b $[Cu_2(18)_4][ClO_4]_4$ with an encapsulated perchlorate anion

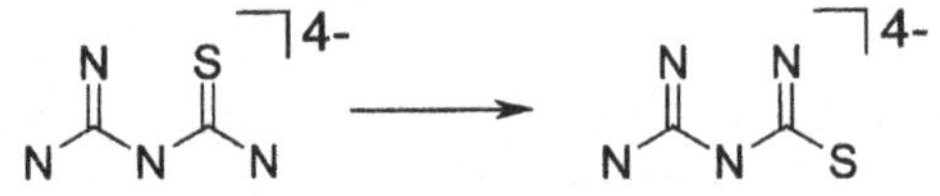

19

Scheme 5 The tautomeric forms of amidinothiourea

X-ray structure of the chloride complex (Fig. 16) the chloride is hydrogen bonded to eight NH groups (two with each ligand at an average NH···Cl distance of 2.43 Å). The Cl^- ion is also close enough to the Ni atoms to interact with them (3.13 Å) and causes a distortion of the square planar Ni geometry. The bromide complex displays a similar structure. Cage formation is not observed when nitrate, acetate or perchlorate are used as the counter-anions, although when $NiCl_2$ is added to these solutions the cage is observed to assemble. This strongly suggests that the cage formation is templated by correctly sized spherical anions.

Non-spherical templating species have been adopted for use in other systems. The tetrahedral BF_4^- has been used for systems such as the Ni square reported by Dunbar et al. [56]. In this process, $[Ni(CH_3CN)_6][BF_4]_2$ self-assembles with

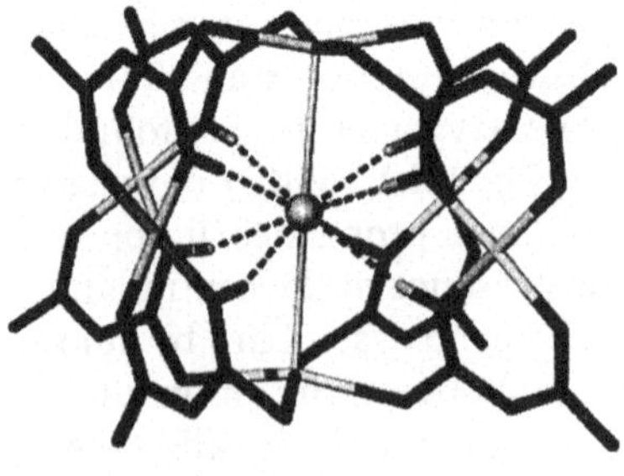

Fig. 16 X-ray crystal structure of $[Ni_2(19)_4]Cl_2$ with the central chloride atom hydrogen bonding to NH groups within the ligands and distorting the geometry of the Ni centre

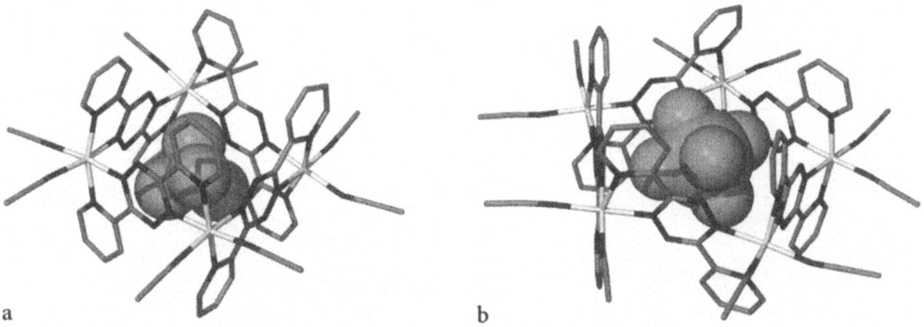

a b

Fig. 17a,b The crystal structures of: **a** $[Ni_4(\mathbf{20})_4(CH_3CN)_8][BF_4]_8$, templated around the central anion; **b** the $[Ni_5(\mathbf{20})_5(CH_3CN)_{10}][SbF_6]_{10}$ pentagon, non encapsulated anions and hydrogen atoms are not shown

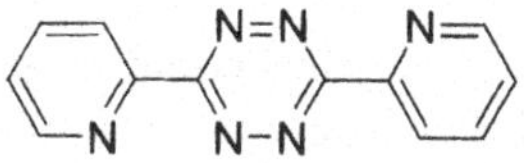

20

3,6-bis(2-pyridyl)-1,2,4,5-tetrazine, **20**, to yield a molecular square with a BF_4^- ion in the centre (Fig. 17a). The square is built upon octahedral Ni centres with two bidentate ligand contacts and two acetonitrile groups on each. A similar structure is also seen when perchlorate is used as the counter anion. The use of SbF_6^-, a larger anion, results in the formation of a unique molecular pentagon (Fig. 17b) [57]. This clearly shows the effect that anion size has in the templated assembly of these cages.

The tetrafluoroborate anion has also been used to template tetrahedral cages [58, 59]. The vast majority of metal cage compounds use rigid ligands, so as to minimise the number of potential products and to ensure a predetermined preference of the ligand interaction for a certain metal coordination geometry. The anion templated systems by McCleverty, Ward et al. use the related flexible bis-bidentate ligands **21** and **22**, based on pyrazolyl-pyridine groups. Ligand **21** has been observed in two highly contrasting, metal dependant structures; tetrahedral $[Co_4(\mathbf{21})_6(BF_4)][BF_4]_7$ (Fig. 18a) and dimeric $[Ni_2(\mathbf{21})_3][BF_4]_4$ (Fig. 18b). Whereas the cobalt structure shows the expected shape from the bis-bidentate ligand, the nickel analogue has only one bridging ligand with the other two acting as tetradentate terminal ligands for the two nickel atoms.

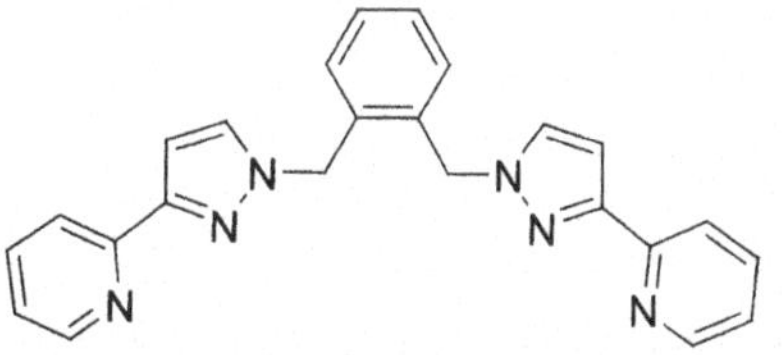

21

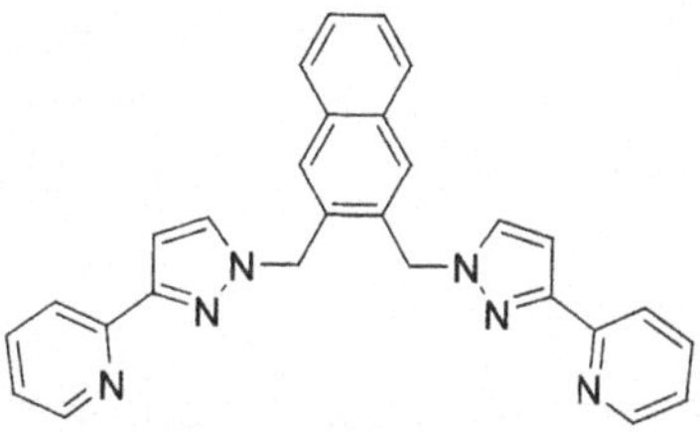

22

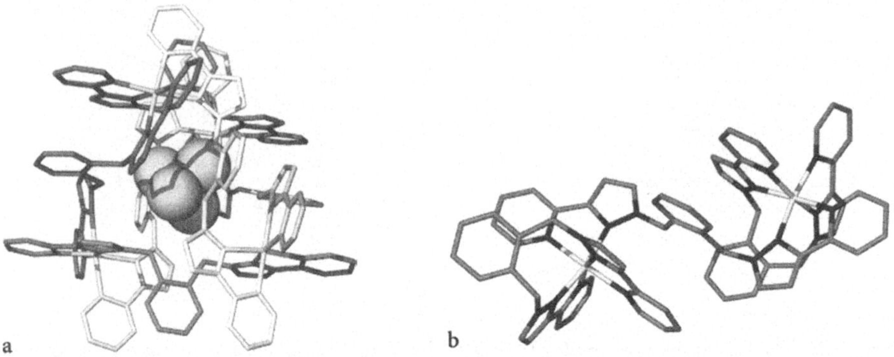

Fig. 18 **a** The tetrahedral Co and **b** bridged Ni compounds of **21**, non-encapsulated anions and hydrogen atoms omitted for clarity

The BF_4^- anion within the tetrahedral cobalt cage is a highly complementary guest in terms of size and shape. There are no interactions between the metal atoms and the guest but a large number of C-H···F hydrogen bonds exist (Fig. 19). There is no exchange of the encapsulated anion with any of the remaining BF_4^- units outside of the cage on the NMR timescale. NMR experiments proved that a templating effect occurs by the addition of one equivalent of BF_4^- to a mixture of cobalt acetate and **21**. Upon this addition the cage was observed to form quantitatively. Other anions, PF_6^- and ClO_4^-, were also tested in this manner and, as expected, perchlorate also displayed a templating effect, as it is of the same geometry as BF_4^-. The analogous compounds of ligand **22** were also shown to exist and display anion templated assembly.

Stang et al. have recently reported a solution equilibrium between trigonal M_3L_3 and square M_4L_4 structures than can be controlled by the ratios of different anions present (Fig. 20) [60]. A simple *trans*-bis(4-pyridyl)ethylene ligand was dissolved with the protected metal complex, [*cis*-$(Me_3P)_2Pt(CF_3SO_3)_2$] in ni-

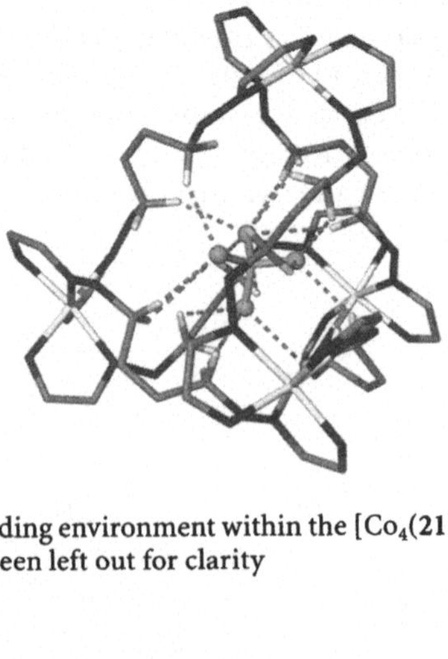

Fig. 19 The hydrogen bonding environment within the [$Co_4(\mathbf{21})_6(BF_4)$] cage. Aromatic groups around the ligands have been left out for clarity

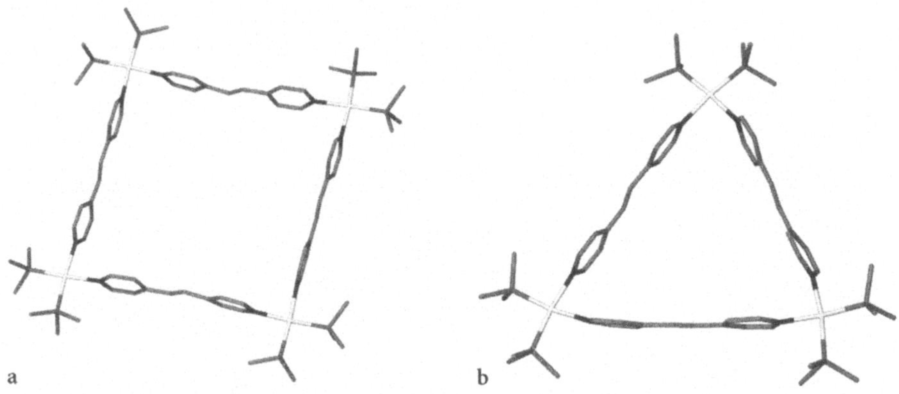

Fig. 20 a M_4L_4 and b M_3L_3 species observed in an equilibrium that can be effected by the anions present, anions not shown for clarity

tromethane. When only triflate is present as the counter anion the molecular square crystallises preferentially, but when two of the triflate anions are exchanged for cobalticarboranes the triangular form is preferred. It is thought that the large steric bulk of the cobalticarborane anion has an effect on the equilibrium of the system.

3.3 Assembly via Panelling Ligands

The use of 1D 'scaffolding' ligands forms polyhedra by joining together the edges of the polyhedra, leaving the faces open. 2D ligands assemble by the coming together of faces of the polyhedra, leading to more enclosed cages. The ligands used are often planar or contain a small curvature which leads to a slightly convex, pseudo-spherical cavity. This method of host assembly has been termed 'molecular panelling'.

The ligand **23** and related species have been used by Fujita et al. to form a variety of cages capable of stabilising a number of unstable species [61]. This trigonal shape is ideal for the construction of an array of polyhedra such as tetrahedra, hexahedra and octahedra. When ligand **23** is stirred in solution with $[Pd(NO_3)_2(en)]$ the octahedral structure **24** is produced, in which alternate faces are occupied by the ligands **23** which replace the nitrate anions (Fig. 21). The en group acts as a blocking unit to force a *cis* coordination geometry for the pyridyl ligands. It is the fact that half of the faces of the octahedron are unoccupied by ligands that provides the capsule with interesting host-guest chemistry.

The octahedron **24** has a very large internal cavity, capable of enclosing multiple neutral guests [62]. Guests have included carboranes, adamantane, adamantols and 1,3,5-trimethoxybenzene. In all of these cases a host:guest ratio of 1:4 is observed with no lower ratios. The hydrophobic cavity that the cage possesses promotes the encapsulation of the above guests in D_2O solution. Tri(*tert*-butyl)benzene was only observed to be encapsulated in a 1:1 ratio. The size of this

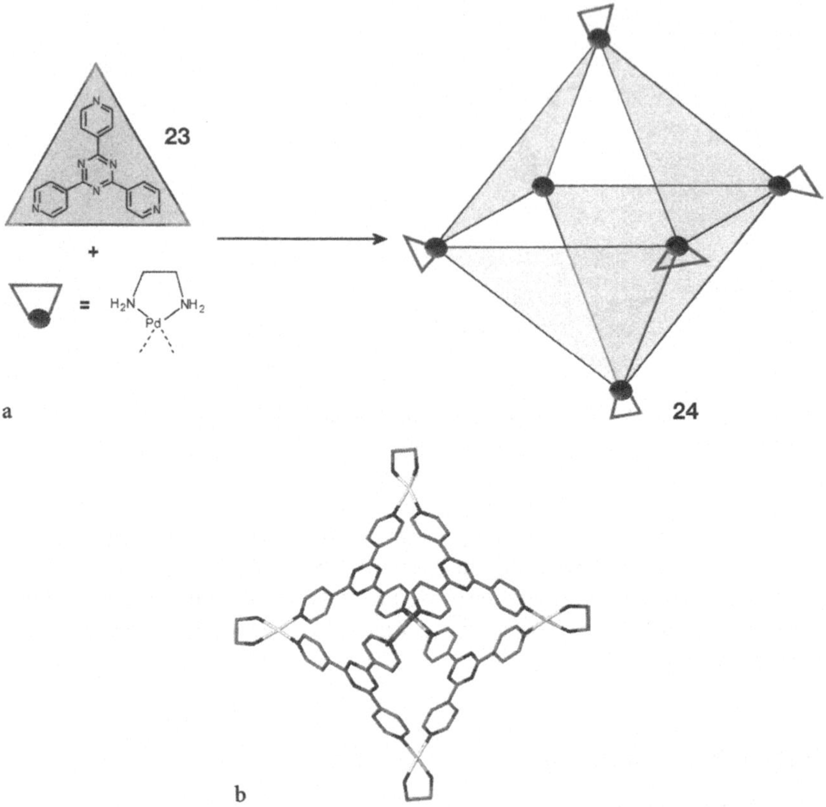

Fig. 21 **a** The formation of the octahedral cage **24** with alternately filled faces. **b** The X-ray crystal structure of an empty host molecule

guest is significantly greater than the others and is larger than the holes presented in the cage. After 2 h at 80 °C the 1:1 complex is observed in a 40% yield. The proposed mechanism of guest encapsulation is by thermally induced slippage. The encapsulation of adamantanecarboxylate in the X-ray crystal structure clearly shows the hydrophobic influence of the binding process, with the hydrophobic adamantane core within the host and the hydrophilic carboxylate functionality directed outwards through the empty cage faces (Fig. 22) [63].

One of the more remarkable encapsulation phenomena exhibited by **24** is the formation of guest dimers within the capsule by a 'ship-in-a-bottle' type process [64]. *cis*-Azobenzene and *cis*-stilbene derivatives were observed to non-covalently dimerise within the cavity. A solution of **24** in D_2O was stirred with a *cis*/*trans* mixture of the potential guest in hexane and after half an hour at room temperature the ^{1}H-NMR spectrum of the D_2O layer showed the encapsulation of only the *cis* isomer in a 1:2 host:guest ratio. No conversion of the guests to the more stable *trans* isomer was observed despite weeks spent in daylight. Molecular modelling has shown that the stabilisation of the guests is due to the formation of a hydrophobic dimer within the cavity (Scheme 6). This dimer is too large

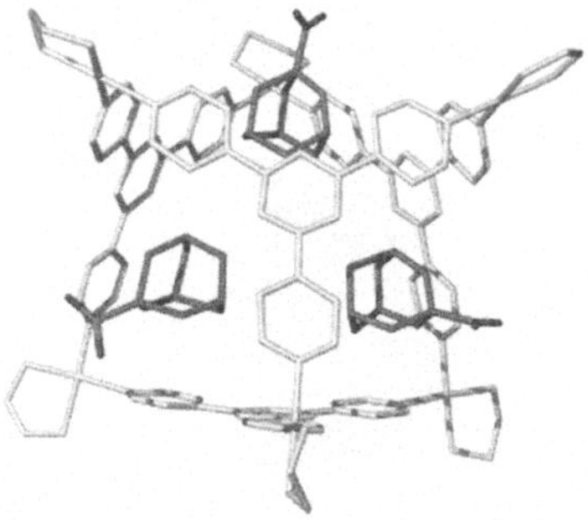

Fig. 22 Crystal structure of the **24** complex with adamantanecarboxylate showing the arrangement of hydrophobic and hydrophilic parts of the guest

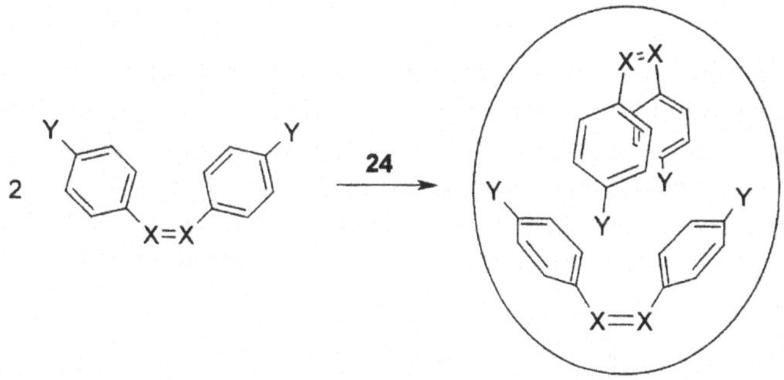

Scheme 6 Derivatives of azobenzene (X=N) and stilbene (X=CH) with methyl or methoxy (in the case of stilbene) Y groups are observed to dimerise within the cavity of **24**

to form outside the cavity and then be encapsulated. The only way in which this species could come to exist is if the monomers diffuse into the cavity sequentially and the dimer forms in situ, in much the same way as a model ship is assembled within a bottle. This unusual arrangement has been observed experimentally by the existence of strong NOE contacts between the methyl groups and aryl CH protons, which do not exist in the spectra of the free guest. X-ray crystal studies showed that 4,4-dimethoxybenzoyl, **25**, exists as an analogous dimeric aggregate inside the cavity of the cage in a chiral, twisted conformation (Fig. 23) [65].

Host **24** also stabilises highly reactive cyclic silanol oligomers by a similar in situ synthesis [66]. Phenyltrimethoxysilane, **26**, rapidly forms polymeric material in aqueous solutions. The presence of **24** prevents this from occurring and isolates the cyclic trimer intermediate, **27**. Several molecules of the silane starting

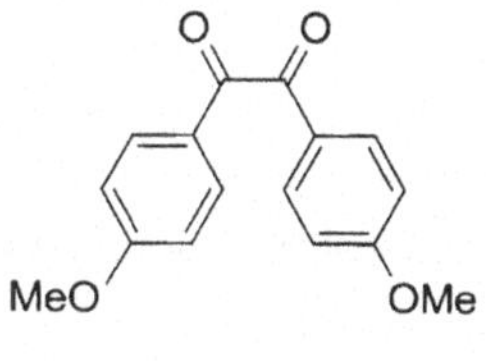

25

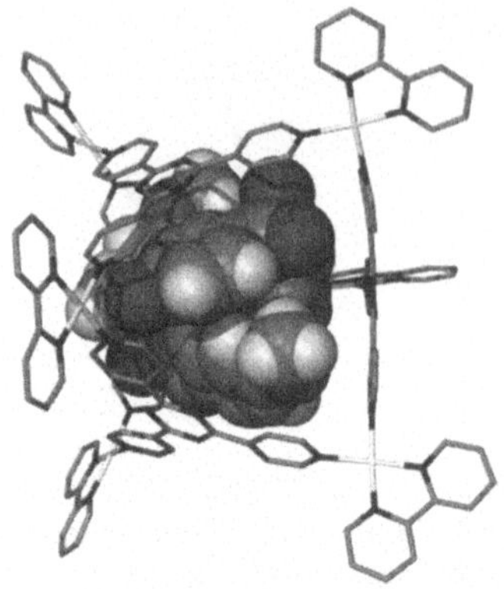

Fig. 23 The crystal structure of a cage of type **24** encapsulating 2 molecules of *cis*-**25** as a dimer

material enter the cavity where they are hydrolysed to the intermediate triol, before undergoing a condensation reaction (Scheme 7). This reaction is prevented from continuing past **27** due to the steric confines of the cage. The nature of the final host-guest complex can be accurately elucidated by ^{1}H-NMR, where the highly symmetric spectrum of the free host is altered by the nature of the guest. The structure was later confirmed by single crystal X-ray analysis (Fig. 24) [67].

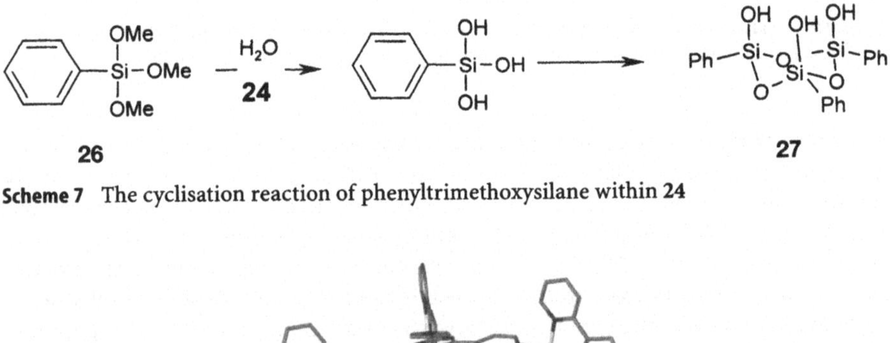

26 **27**

Scheme 7 The cyclisation reaction of phenyltrimethoxysilane within **24**

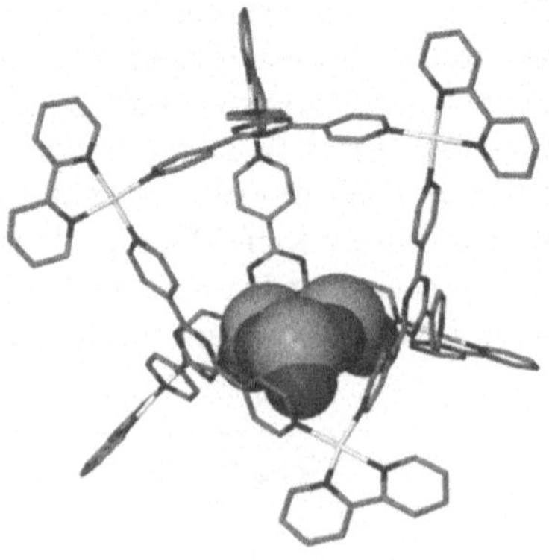

Fig. 24 X-ray structure of **27** protected within a cage of type **24**, disordered phenyl groups not shown for clarity

The most recent phenomenon observed to occur within the confines of **24** is the stereoselective photodimerisation of olefins [68]. Without the presence of the cage, the dimerisation of **28** results in a mixture of the syn- and anti- isomers (**29** and **30** respectively, Scheme 8). With the cage present, two molecules of **28** are encapsulated before the photodimerisation occurs. Within the cavity the syn isomer is the only one capable of forming for steric reasons. This type of selectivity has been observed for several related compounds. Recently, photodimerisation has also been observed to be controlled by supramolecular templation in the solid state [69].

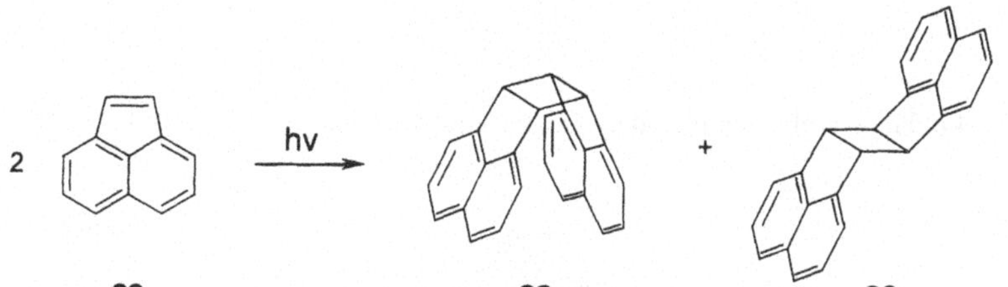

28 **29** **30**

Scheme 8 The photodimerisation reaction which can be controlled by **24**

Other structural motifs have also been synthesised that rely on similar pyridine-palladium interactions to assemble panelled polyhedra. A molecular tube, almost cuboidal in shape has been prepared using **31** [70]. When combined with $[Pd(en)(NO_3)_2]$ the tube-like structure formed, but only when templated by the presence of the 4,4′-biphenylenedicarboxylate dianion, which has been shown by X-ray crystallography to reside within the cavity (Fig. 25). The way in which the ligands are arranged at 90° angles to each other means that the aromatic guest can interact with the host via both face-to-face π-stacking and CH-π interactions. This guest inclusion method has also been used in the stabilisation of reactive silane intermediates, complexing the triol **32** and preventing any polymerisation from occurring [67].

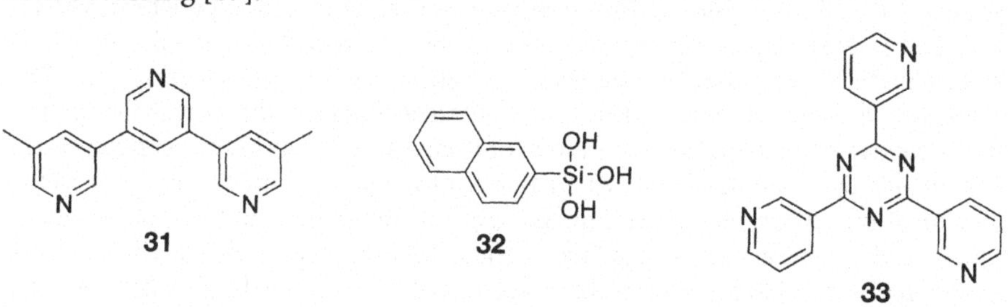

31 **32** **33**

Ligands of type **23** are very versatile, in that the positions of the nitrogen atoms of the pyridine rings can be altered to provide a new ligand. If the *meta* isomer is used instead of the *para*, the new ligand, **33**, displays very different self-assembly behaviour [71]. Instead of an enclosed cage system, a deep bowl is formed (Fig. 26). This assembly stabilises silane dimers within its cavity [67].

The bowl structure is observed to dimerise in the presence of the correct guest [72]. The bowl was studied using *o*-, *m*- and *p*-terphenyl as potential guest

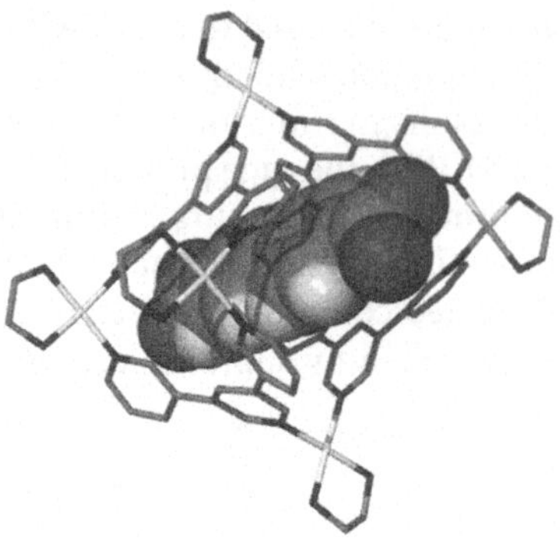

Fig. 25 Cuboidal assembly templated by an aromatic dianion

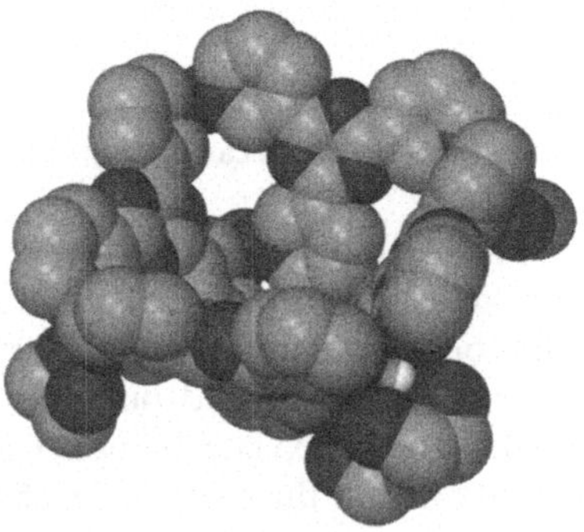

Fig. 26 X-ray crystal structure showing the geometry of the $[Pd(en)]_6[\mathbf{33}]_4$ bowl, hydrogen atoms are omitted for clarity

species. With all these guests, dimers were observed with a host:guest ratio of 1:2 (2:4). In the case of *o*-terphenyl the dimers were formed by hydrophobic interactions bringing together two bowls, each containing two guests (Fig. 27a). The *m*-terphenyl dimer showed a different structure, whereby the guests bridge between the two bowl structures, with two of their rings in one hemisphere and one in the other, involved in CH-π hydrogen bonding via the aromatic rings with the interior of the cavity (Fig. 27b). The linear *p*-terphenyl has not been observed as a guest as the geometry does not favour encapsulation. Hydrophobic dimers have also been observed to form with six molecules of *cis*-stilbene divided between the two hemispheres.

Further adaptation of **23** has been shown to lead to other geometries such as tetrahedra and hexahedra [73]. For example, **34** leads to the assembly of the ligands in an edge-to-edge manner, unlike the corner-to-corner motif adopted in the octahedral cage **24**. These polyhedra have all of their faces occupied by ligands and provide a very enclosed environment (Fig. 28). The hexahedral structure displays guest exchange properties with a variety of methyl halides [74]. This can occur through the small gaps left at the nonbinding sites of the ligand, where

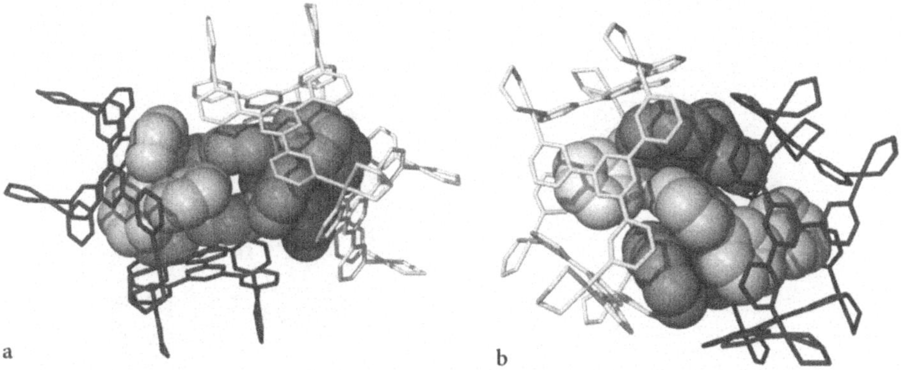

Fig. 27a, b X-ray crystal structures of hydrophobic dimers of the bowl **33** with: **a** *o*-terphenyl; **b** *m*-terphenyl as guest species

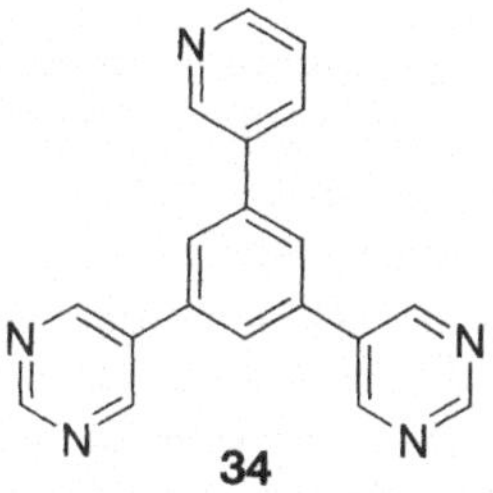

34

there is only one nitrogen atom on the ligand terminus. Guests such as tetrabromomethane have been encapsulated in a 1:2 host:guest ratio in a hydrophobically driven processes within 10 min at room temperature. The guest is quickly decapsulated upon addition of ethanol to the D_2O solution. Related ligands have also shown heteroassociation from a dynamic library which can be influenced by the guest species present [75].

The above ligands are all rigid triangles with very little flexibility. However, Fujita and co-workers have developed an M_3L_2 cage-like complex by using the

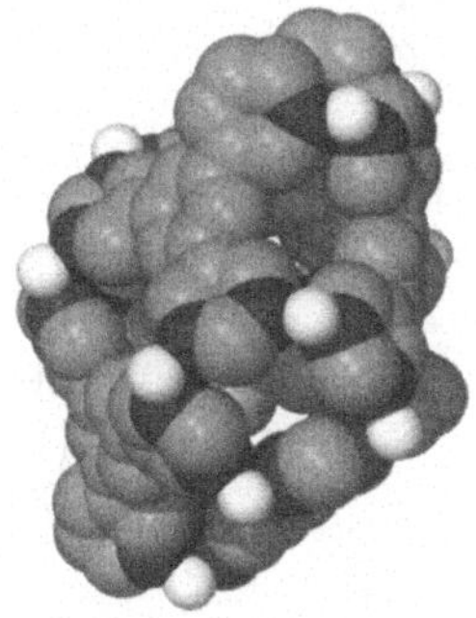

Fig. 28 X-ray crystal structure of a molecular hexahedra constructed using **34** and Pd(en) units. En groups and hydrogen atoms not shown for clarity. Pd atoms modelled as pale spheres

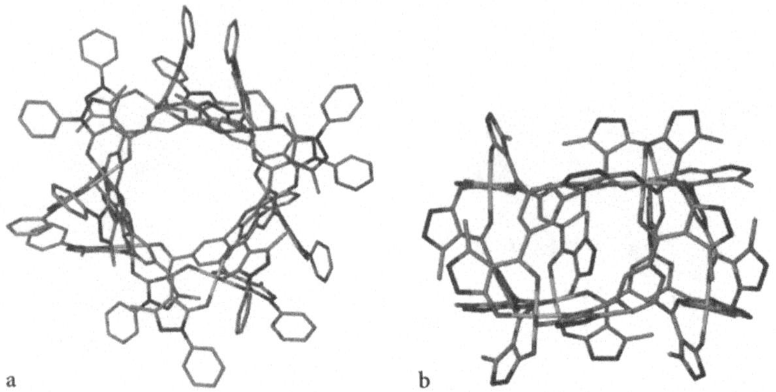

Fig. 29a, b $Ga_6(35)_6$ cylindrical cage viewed: **a** from above; **b** from the side with phenyl groups omitted for clarity

more flexible ligand 1,3,5-tris(4-pyridylmethyl)benzene with (en)$Pd(NO_3)_2$. This complex assembles in high yields only in the presence of specific guest molecules containing bulky hydrophobic moieties such as 1-phenylethyl or adamantyl groups. In the absence of a suitable guest, only oligomeric products are obtained [76]. Similar trigonal ligands have also been observed to form cage complexes. The ligand **35** has been utilised by Raymond and co-workers to form an M_6L_6 cylindrical cage with a substantial cavity, open at either end [77]. The flexible nature of this ligand allows a cylindrical shape to assemble around six octahedral gallium atoms with the bidentate β-ketone arms. The crystal structure of the cage (Fig. 29) shows that there is disordered water within the cavity, meaning that it should be possible for the cage to show host behaviour for other species although to date none has been reported.

Sun et al. have used the flexible trigonal ligand **36** to form a smaller M_3L_2 cage compound with zinc acetate [78]. This system represents one of the first such compounds to be formed and subsequently analysed in the solid state. The cage was shown to complex camphor with a binding constant of 117 M^{-1} (D_2O). Fur-

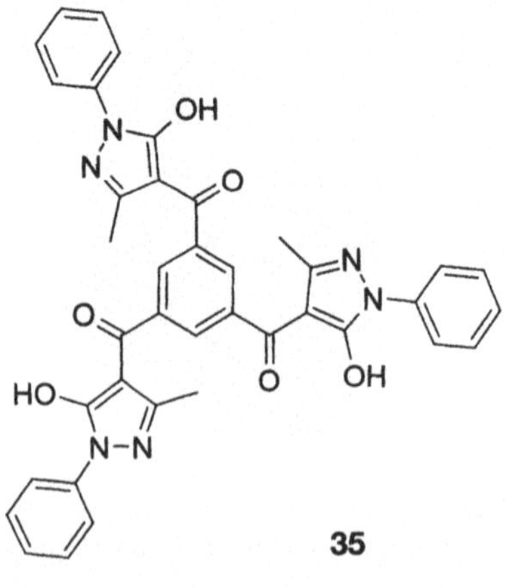

35

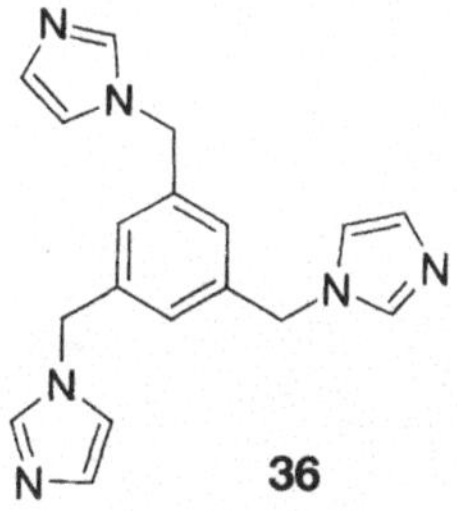

36

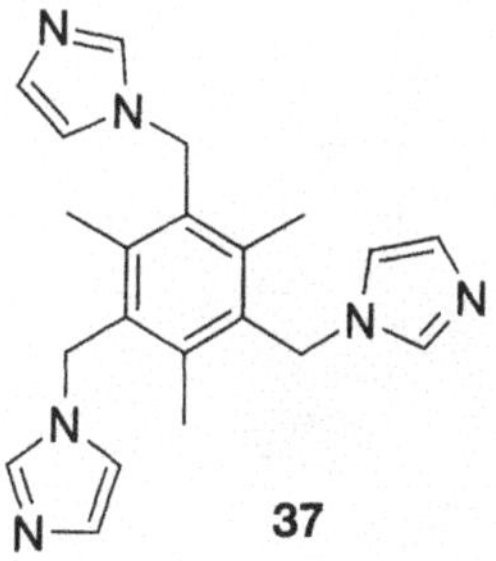

37

ther work was carried out using the modified ligand **37** with silver as the coordinating metal [79]. The $Ag_3(\mathbf{37})_2$ cages were formed using either silver perchlorate or silver tetrafluoroborate in ethanol. The compounds formed from these two different reactants were found to be isostructural. Solid state structures of the perchlorate complex show an anion situated in the centre of the cavity (Fig. 30a). When the same reactions were carried out using acetonitrile and dimethylformamide as solvents a different product was obtained. X-ray analysis of this new compound showed it to have the same connectivity but distorted, with the coordinating anion existing on the edge of the cavity (Fig. 30b).

The crystallographic studies led the way into studying the exchange properties of the host; if the anion can exist between two ligand arms then it should be able to pass through them as part of an exchange process. It was found that BF_4^- could be displaced by ClO_4- when the tetrafluoroborate complex was placed in an excess amount of $NaClO_4$. No kinetic data have been established for this process.

Steel and co-workers have reported the formation and X-ray structure of an M_6L_4 cage comprised of six *trans*-dichloropalladium units bridged by four molecules of the mesitylene-derived ligand **38** (Fig. 31). The six palladium atoms are arranged in a pseudo-octahedral array, while the four molecules of **38** form a tetrahedrally disposed internal core of benzene rings with an internal radius of approximately 4.7 Å. Within this cavity resides a single, disordered $(CH_3)_2SO$ molecule. The cage is maintained in solution. Although constitutionally similar to the cage **24**, this structure is topologically quite different, a consequence of the different coordination geometries of the palladium atoms.

More complex shapes can also be constructed using simple polygonal ligands. Stang and co-workers have used triangular ligands containing aromatic rings and

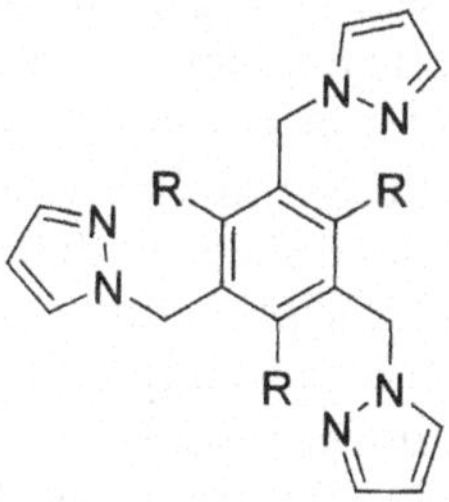

38

R = H, Me, Et

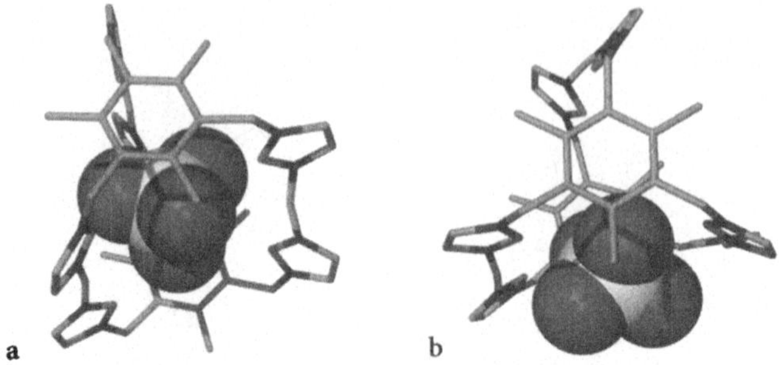

Fig. 30a, b The two X-ray crystal structures of $Ag_3(\mathbf{37})_2$ with the perchlorate anion located in different positions: **a** inside the cavity; **b** part-way inside the host

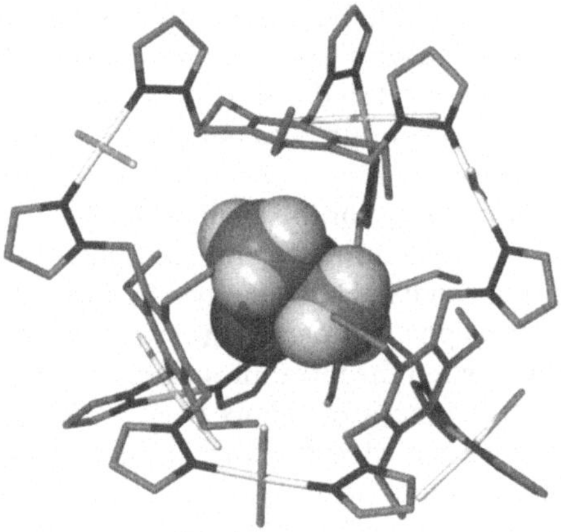

Fig. 31 The X-ray structure of the M_6L_4 complex between ligand **38** and *trans*-palladium chloride units with an encapsulated molecule of DMSO

alkyne groups which provide long rigid arms. Such ligands, when combined with linear bidentate ligands, give rise to cuboctahedral architectures [80]. These shapes can be constructed via a 3:2 mixture of bidentate and tridentate ligands, using 20 components in total to close the cage. Two methods of synthesising this complex structure have been developed (Scheme 9). These alter in the placement of the platinum metal before the assembling of the cage, although once the cage is formed the geometry is essentially the same whichever route is taken. Both assembly processes occur within 10 min at room temperature. Similar methods have also been employed to synthesise dodecahedral cages [81]. The ligands utilised in this case are of a pyramidal shape with three coordinating arms and straight bidentate ligands to which the platinum metal centres are attached.

Pyramidal, tridentate ligands have recently been utilised to synthesise very wide cages when combined with linear ligands to form $M_6L^t{}_2L^b{}_3$ (L^t=tridentate ligand, L^b=bidentate ligands) [82]. The tridentate ligands have been centred around either a carbon atom, **44**, utilising its tetrahedral geometry, or around an adamantane core, **43**. As with the synthesis of the cuboctahedra, the metal atoms

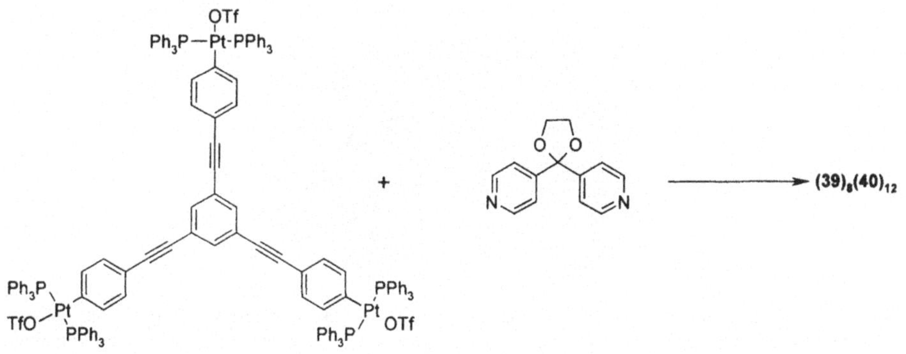

39 40

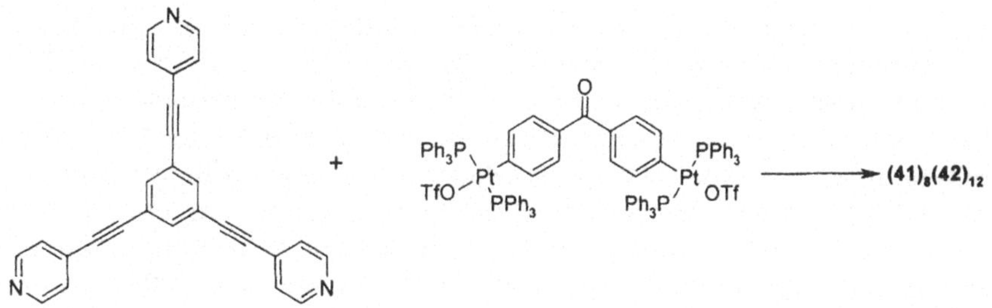

41 42

Scheme 9 Two methods for the synthesis of a cuboctahedral cage from a mixture of tridentate and bidentate ligands (OTf=Triflate). Final cages carry a 24$^+$ charge with triflate counter anions

are incorporated into one of the ligands, in this case the linear ligand derived from **45**, to reduce the possible species that could assemble. Both of the tridentate ligands, **43** and **44**, when combined with **45**, form wide capsules ranging from 2 to 4 nanometres across.

Single crystal X-ray analysis of the cage formed from **44** and **45** showed that one of the nitrate counter anions exists within the cavity (Fig. 32). This anion fits exactly within this intra-capsule space, with the oxygen atoms pointing into the gaps in the sides of the cage between the bidentate ligands. This cage is observed to form an order of magnitude faster than other similar cages, showing some evidence of a templating effect by the anion. No exchange of this guest has been

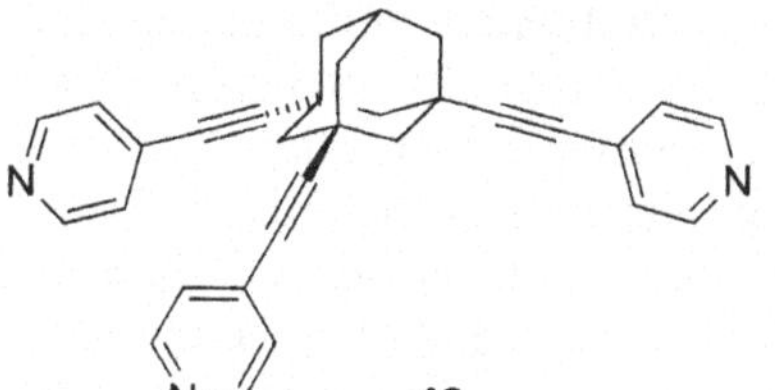

43

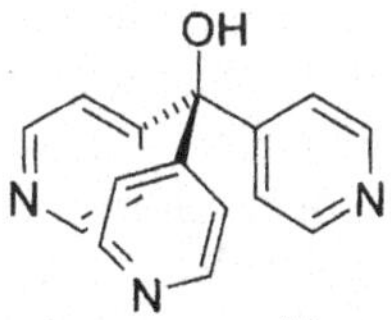

44

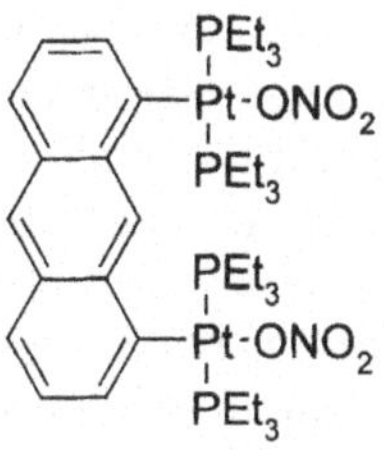

45

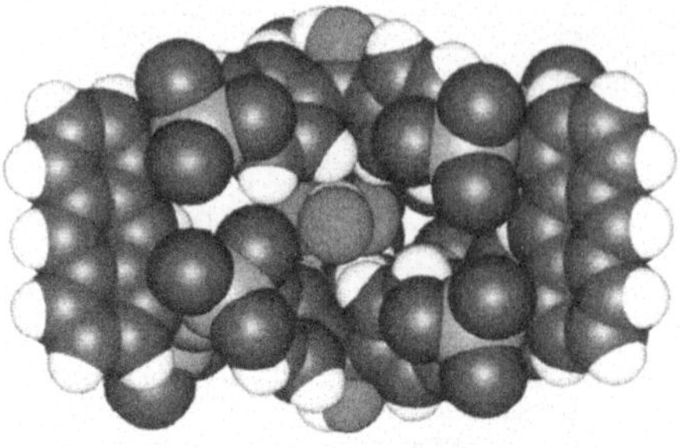

Fig. 32 X-ray crystal structure of $(\mathbf{44})_2(\mathbf{45})_3$ cage with the encapsulated nitrate anion visible, orientated so that the oxygen atoms face through gaps in the cage, reproduced with permission from [82], copyright 2002 National Academy of Sciences, USA

observed, even when a large excess of KPF_6 is added. The adamantane-based cage is believed to have a similar structure by virtue of similarities in its ^{1}H-NMR spectra with cages whose structures have been confirmed in the solid state. Stang has also used the tridentate ligand **44** to prepare a self-assembled M_3L_2 trigonal-bipyramidal supramolecular species, **46**, by combination with three ditopic platinum acceptors [83]. Lehn and co-workers have prepared cages based on a combination of bi- and tridentate ligands [84]. In contrast to the work by Stang, these systems assemble around naked metal ions and have been extended to form compartmentalised helices [85].

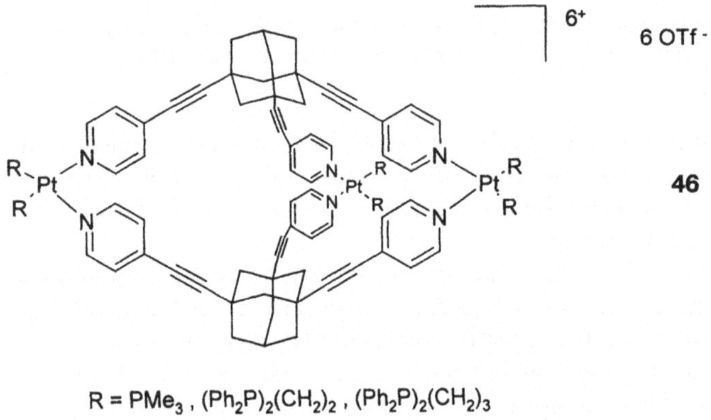

46

Although triangular ligands are the most widely used in the formation of coordination cages, other shapes have also been investigated. Fujita has used a square tetradentate ligand which forms either a box-like M_6L_3 structure or an M_8L_4 trigonal pyramid [61]. A similar trigonal pyramidal structure has been reported using tetradentate zinc-porphyrin complexes as the ligands. Two porphyrin-based building blocks intermolecularly bind with four *cis*-Pd(II) complexes resulting in a novel molecular capsule, **47**, with a cavity large enough to hold bipyridine as a guest. A K_{ass} of 2.6×10^6 M^{-1} has been measured for this guest in $CHCl_3$ [86].

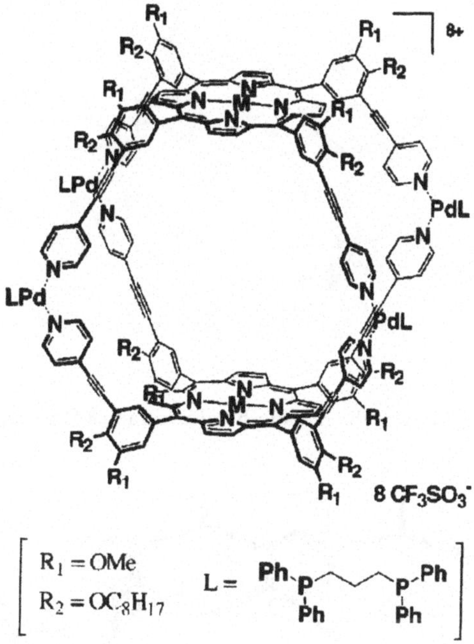

47

Apart from ligands that are flat or slightly distorted polygons, there has been an increasing trend in studying bowl-shaped ligands which can be joined to give capsules very similar to hemicarcerands. The first fully characterised example of this type of system was reported by Harrison et al. [87]. Two resorcinarene molecules **48** with four tridentate ligands around the rim are joined together by four cobalt (II) ions, as shown in the crystal structure (Fig. 33). The resulting cage **49**

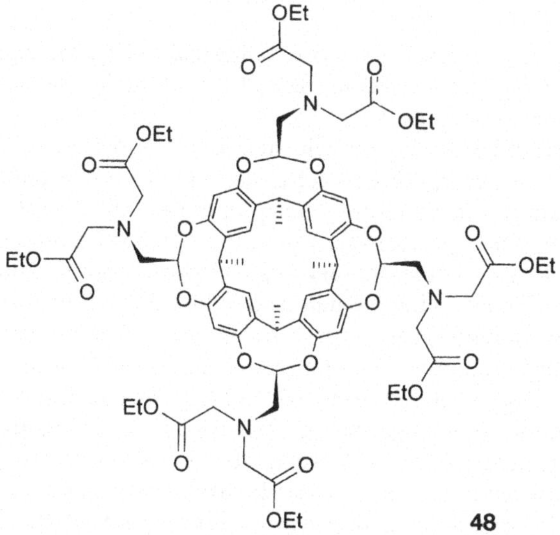

48

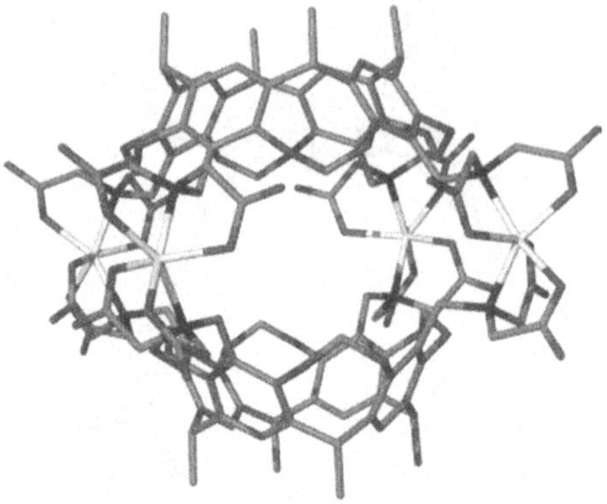

Fig. 33 X-ray crystal structure of the metallo-hemicarcerand **49** developed by Harrison et al.

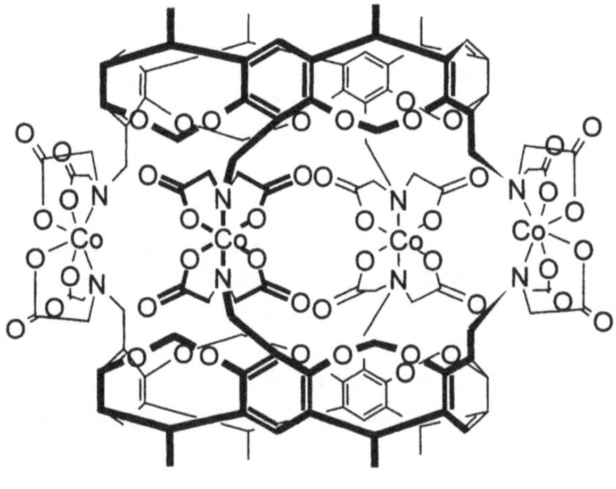

49

is octaanionic and possesses a roughly spherical cavity. Molecular modelling shows that the cavity is of a suitable size to fit small molecules such as benzene, acetone and dichloromethane.

Many systems similar to this metallo-hemicarcerand exist, differing predominantly in the metals and ligating moieties used. For example, palladium systems exist which are similar in structure to cryptophanes [88]. The choice of metal is usually limited to either palladium or platinum. This is because the square planar geometry offers a complementary angle between *cis*-coordination sites and the resorcinarene ligands. These metals also reduce dimensionality, favouring the formation of non-polymeric structures. Dalcanale et al. have used nitrile groups positioned around the upper rim of a resorcinarene to coordinate to platinum [89]. The C-N-Pt angles ranged from 165° to 172°. The coordination geometry of the metal was found to be of great consequence as the analogous nickel(II) system, which exists in equilibrium between tetrahedral and square planar geometries, was not observed to facilitate capsule formation. Cages of a similar geometry have also been produced that use pyridyl ligands [90]. The strength of metal-pyridyl interactions is greater than that of nitriles enabling studies to be carried out under harsher conditions, such as in DMSO which can compete as a ligand. The greater inertness of platinum coordination is also evident, with the platinum cage being more kinetically stable than the analogous palladium cage. Similar cages utilising pyridyl ligands have been shown to trap cationic guests during formation [91] and pyridyl ligands have also been used to form cages from calixarenes [92–95].

More complex architectures can also be assembled via the coordination of resorcinarenes to metals, such as the resorcinarene based loop and tetrahedron reported by Beer [96]. The resorcinarene **50** is functionalised with four dithiocarbamate moieties which can coordinate in a bidentate manner with a variety of metals. The more conformationally flexible 'arms' of this ligand allow for structures other than dimeric cages to be formed. The product of the reaction of zinc acetate and **50** in water/pyridine is a roughly triangular molecular loop, $[Zn_6(\mathbf{50})_3(py)_6]$. Each point of the triangle is occupied by a resorcinarene bowl with two dithiocarbamate-Zn(py) bridges forming the edges. Space-filling CPK representations based on the crystal structures show that the cavity is circular with an approximate diameter of 16–17 Å and therefore suitable for the binding of spherical guests, such as C_{60} (Fig. 34). The trimer assembled by six cadmium or zinc ions was shown by UV-vis titrations to strongly bind C_{60} in toluene and benzene solutions [97].

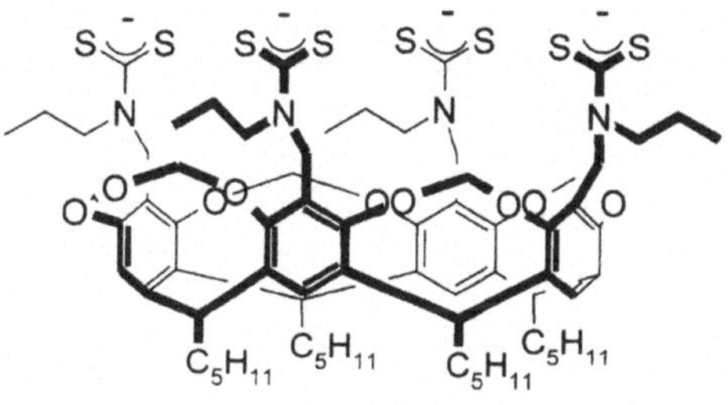

50

The resorcinarene/dithiocarbamate system also displays a marked dependence on the choice of metal used. When copper acetate is combined with **50** a completely different structure is formed; a distorted tetrahedron containing four resorcinarenes and eight copper atoms. As with the triangular zinc-based structure, the resorcinarenes are located at the corners of the polyhedron with metal bridges joining them together. To date no guest inclusion studies have been published for this unique compound but it is conjectured that the large portals within the tetrahedral cage could allow for guests to enter the cavity and that the redox-active nature of the metal bridges could lead to sensing applications. Smaller macrocycles containing Cu-dithiocarbamate bridges have already been reported to show signalling behaviour [98].

Kimura and co-workers have prepared a novel self-assembled cage composed of four trimeric Zn(II)-cyclen complexes joined with four tri-deprotonated trithiocyanuric acid molecules in aqueous solution. The 4:4 complex is thermo-

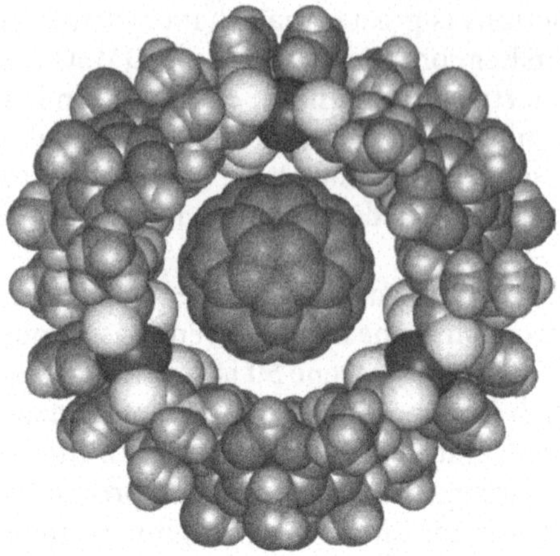

Fig. 34 Space filling representation of $[Zn_6(\mathbf{50})_3(py)_6]$ with a C_{60} fullerene inside the cavity, [97]. Reproduced by permission of The Royal Society of Chemistry

dynamically stable and forms quantitatively through the formation of $Zn^{2+}S^-$ exocyclic bonds. The exterior may be viewed as a cuboctahedral architecture possessing a discrete nanoscale inner cavity (a truncated tetrahedral cage) able to encapsulate size-matching, hydrophobic guests such as 1-adamantanecarboxylic acid, 2,4-dinitrophenol or 7-diethylaminocoumarin-3-carboxylic acid (Fig. 35). These complexes are kinetically stable, however, the guest molecules inside the cavity can be exchanged [99].

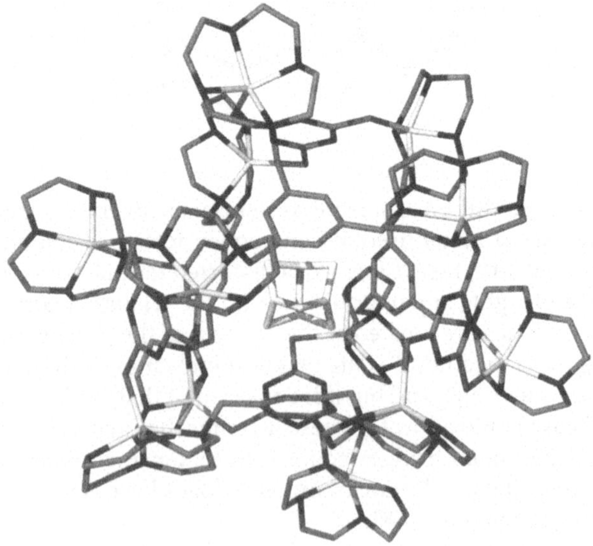

Fig. 35 Admantane contained within a cage formed from four trimeric Zn(II)-cyclen complexes joined with four tri-deprotonated trithiocyanuric acid molecules

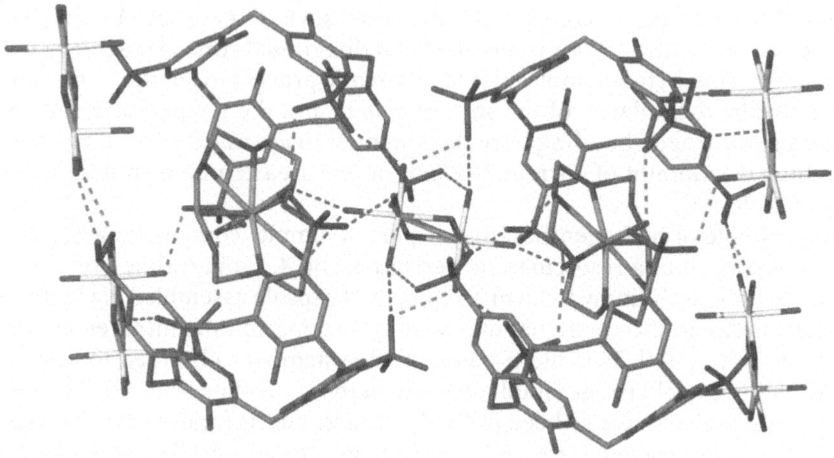

Fig. 36 A cage held together by both metal-ligand interactions and hydrogen bonds

A metal-assembled capsule has been designed by Raston that utilises hydrogen bonding to hold the cage together, as well as the metal-ligand interactions. The X-ray structure shows the formation of a novel multicomponent capsule in which two *p*-sulphonatocalix[4]arene subunits encapsulating an 18-crown[6] guest are linked by the sulphonate groups of the calixarenes by yttrium, europium or rhodium cations (Fig. 36) [100, 101].

4 Non-Covalent Assemblies

Although cavity-containing species are readily accessible via covalent synthesis and coordination interactions, there is an increasing trend towards forming capsules by the non-covalent self-assembly of multiple components. Within solution such assemblies exist in an equilibrium between the capsule and the individual parts. The entry and exit of guests can take place as the capsule is continually assembling and disassembling, with the guest exchange limited by the rate at which the disassociation of the capsule occurs. The guests in such species can be fully enclosed with no contact with the external environment as it is not necessary for portals to be left for exchange. The lifespan of the complexes can be enhanced by the presence of a suitable guest as a template. The variety of non-covalent capsules can be readily divided into those which are assembled from several different species and those which are composed of two or more self-complimentary molecules.

4.1 Multi-Component Assemblies

One of the interactions most commonly utilised, both by the supramolecular chemist and by nature, is the hydrogen bond [23]. The directionality and relative selectivity offered by this interaction makes it very appealing in most cases. When attempting to develop a complex 3D structure, such as a capsule, by the use of 'non-covalent synthesis', it is imperative that directional control is achieved. Hydrogen bonding between molecules is a dynamic process in solution and can be controlled by the polarity of the solvent system and the temperature. Capsules created by hydrogen bonding have the potential to act as excellent host species and the large amount of current research in this area can be seen in recent reviews [21, 24, 25].

An example of hydrogen-bonding as part of a multi-component assembly is the combination of a resorcinarene derivative and 4,4′-bipyridine prepared by Atwood and MacGillivray to form an elongated capsule assembled via hydrogen bonding between the resorcinarene hydroxyl groups and the nitrogen atoms of the bipyridine [102]. This multi-component system was observed to bind two molecules of nitrobenzene simultaneously in the solid state (Fig. 37). The solid-state of this system has also been utilised to isolate guests for time-resolved spectroscopy of their excited states [103]. Raston and co-workers have described the formation of analogous capsules in which two resorcin[4]arenes are linked by four substituted terpyridines. Solvent interactions in the formation of the capsule determine the size and shape of the guest cavity. The capsules have large internal volumes occupied by four guest molecules; either four disordered toluene molecules or, selectively, two toluene molecules and two diethyl ether molecules ordered within the crystal lattice (Fig. 38). Both capsules are held together by a total of eight N···HO hydrogen bonds [104].

Nitrogen-containing aromatic rings generally represent good acceptors for hydrogen bonds. If these can then be further functionalised to provide more hydrogen bonding sites then a very robust capsule can be produced. An example of this is the capsule **52** formed by 2-aminopyrimidine and the tetracarboxylic acid resorcinarene **51** (Scheme 10) [105]. The complimentary geometries of the carboxyl units and the bridging pyrimidine provide sixteen hydrogen bonds holding the capsule together. As with the bipyridine capsule shown in Fig. 37, the resulting 2-aminopyrimidine capsule is observed to encap-

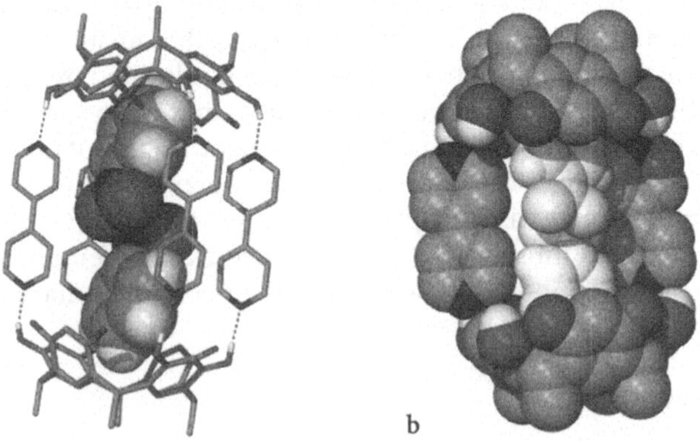

a b

Fig. 37a, b Solid state structures of the resorcinarene-bipyridine system showing: **a** the composition of the capsule; **b** the tight manner in which the guests are bound (two bypyridine groups are omitted to show interior)

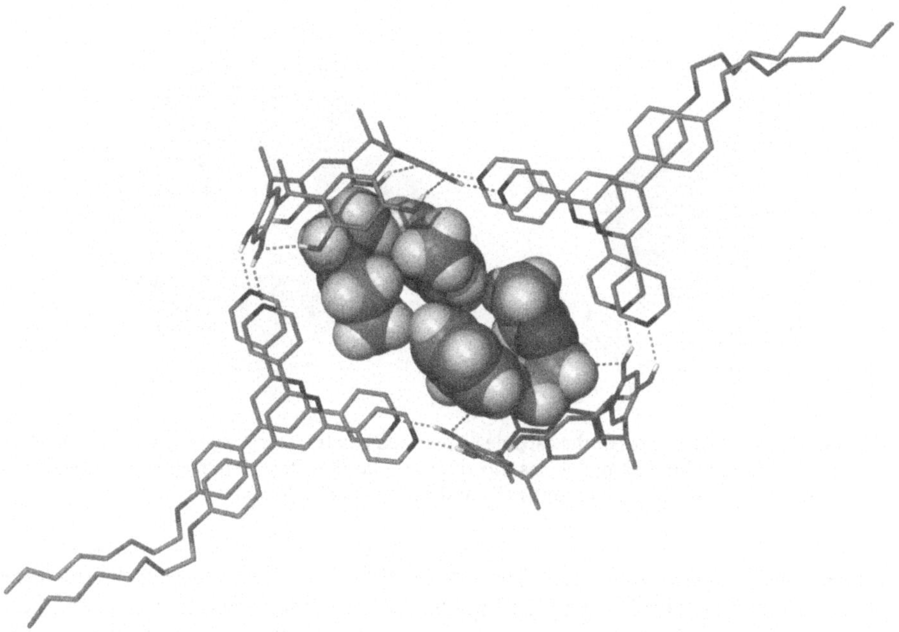

Fig. 38 Crystal structure of the system by Raston and co-workers composed of two resorcin[4]arenes linked by four substituted terpyridines with two toluene and two diethylether molecules encapsulated

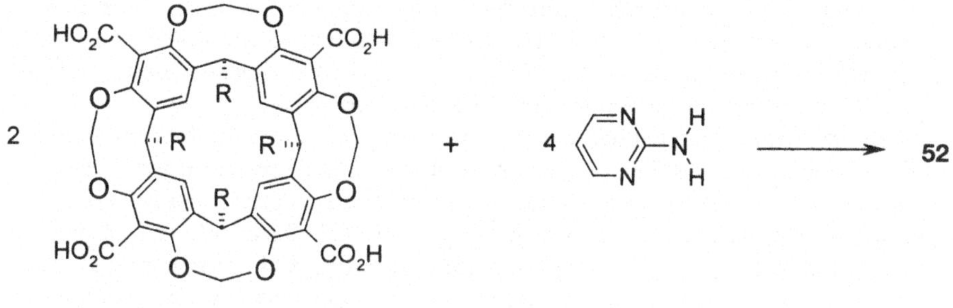

R = $(CH_2)_2Ph$

51

Scheme 10 The formation of capsule 52 (Fig. 39) by the combination of 2-aminopyridine and the resorcinarene 51

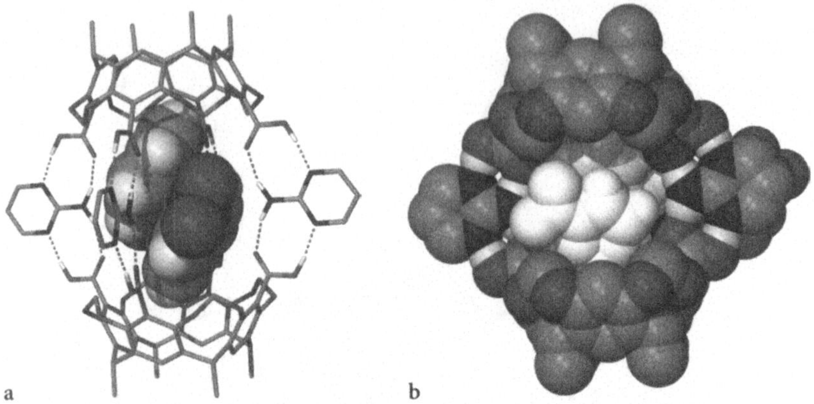

Fig. 39a, b Crystal structures of the capsule **52** between two tetracarboxylic acid resorcinarenes and four 2-aminopyrimidine molecules showing: **a** the composition; **b** the packing of the guests within the cavity (one aminopyrimidine is omitted to show interior)

sulate small, neutral molecules. The crystal structure of the nitrobenzene complex shows a close similarity to the bipyridine system with two guests stacked together, although they are more overlapped due to the shape of the capsule (Fig. 39).

Some other capsules have been described in which both complimentary functions, carboxylic acids and nitrogen-containing aromatic rings, are respectively attached to the rims of two bowl-shaped molecules. Reinhoudt has prepared self-assembling, multi-hydrogen bonding molecular capsules in which a calix[4]arene substituted with four carboxylic functions at the upper rim interacts with a complimentary calix[4]arene with four pyridines attached to the lower rim. These capsules have been identified by ^{1}H-NMR, IR and VPO measurements but encapsulation properties have not been reported [106].

Hydrogen bonded capsules within multi-component assemblies can also include solvent molecules. The first example of this kind of system was reported by Atwood et al. [107]. A resorcinarene with eight hydroxyl functionalities situated around one rim was observed to form roughly spherical assemblies comprising six resorcinarenes and eight water molecules providing 60 hydrogen bonds keeping the sphere intact. The capsule maintains its structure within non-polar solvents such as benzene. The identity of species encapsulated within the structure could not be determined in either solution or the solid state, where unidentifiable electron density was evident. The polarity of solvents used to study assemblies is a very important factor. Highly polar media can compete for, and disrupt, hydrogen bonding networks, making the formation of capsular species impossible. Recently, the first examples of non-covalent assemblies that are stable in competitive solvents have been reported [108–111].

Multi-component, solvent-bridged systems are usually formed between two identical units resulting in a bridged dimer [112–114]. The normal approach for these systems has been the same as that for simple multi-component capsules, namely the use of concave molecules such as calixarenes with hydrogen bonding

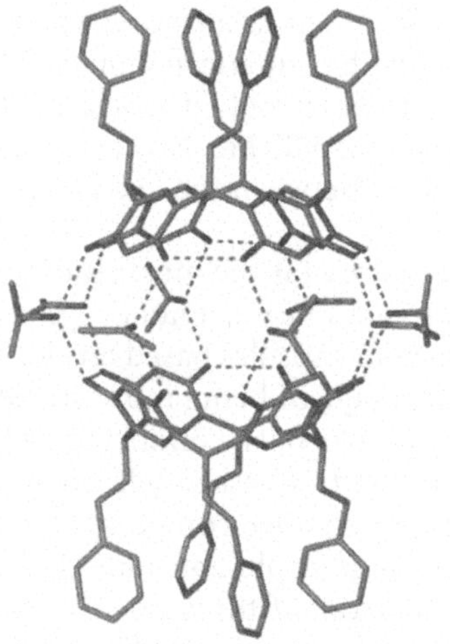

Fig. 40 Two resorcinarene cavitands held together by 16 hydrogen bonds with 8 bridging isopropanol molecules

units situated around one rim. An example of such a system is shown in Fig. 40, whereby isopropanol molecules bridge between two resorcinarene molecules to produce a capsule that is held together by 16 hydrogen bonds [112]. An analogous resorcin[4]arene unit forms molecular capsules in the solid state in which two 'head-to-head' arranged bowls are linked by eight water molecules and encapsulates a tetraethylammonium ion within the cavity via cation-π interactions in a guest-driven assembly [115].

Hydrogen bonding is ubiquitous in Nature and is essential to the most common and most important of systems, such as tertiary protein structures and DNA base pairing. DNA involves highly directed hydrogen bonding between complimentary bases; adenine and thymine (A-T) and cytosine and guanine (C-G). Despite the excellent mutual recognition of these groups very little work has been carried out with them in synthetic systems. Initial work into the pairing of calixarenes using A-T base pairing has been reported by Huang and co-workers (Scheme 11) [116]. The individual adenine-calix[4]arene and thymine-

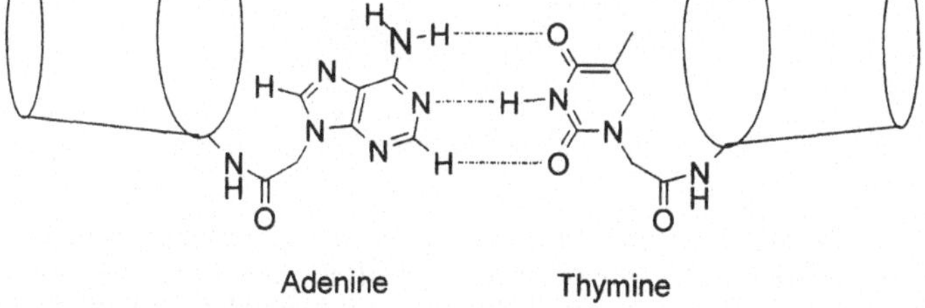

Scheme 11 Two calixarenes joined by Watson-Crick A-T base pairing

calix[4]arene showed significant self-association when alone and when placed in $CDCl_3$ solution together formed a dimeric assembly. When competitive solvents, such as DMSO, were added the assembly dissociated. The pairing was however, stable enough to be observed via ESI-MS. No cavity can be formed in this manner as there is only one bridge between the two calixarenes and at least two are need to close the structure.

Whilst hydrogen bonding is by far the most studied method of forming non-covalent capsules, other interactions must not be ignored. There is a growing interest in the design of molecular capsules based on electrostatic interactions. This type of capsule has the advantage of being more stable in polar, protic solvents. In this sense, pioneering work has been conducted by Schrader and co-workers, who reported simple and versatile access to self-organised spheroidal molecular assemblies composed of highly charged complimentary building blocks based on ammonium or amidinium and phosphonate ions [117–119]. The complex geometry features an alternating array of three positive and three negative charges around the cages central seam, interconnected by a regular network of linear hydrogen bonds (Fig. 41). The association constants reach 10^6 M^{-1} in methanol but span several orders of magnitude reflecting the different degree of preorganisation of the complexation partners used. In water the binding remains strong, in the range of 10^3 M^{-1}.

Positive enthalpy and entropy terms were calculated meaning that, contrary to self-organisation processes in apolar solvents, the enthalpic gain and entropic cost are completely overriden by solvation effects. Attempted inclusion of even small diatomic guests leads to a widening and thus a destabilisation of the cap-

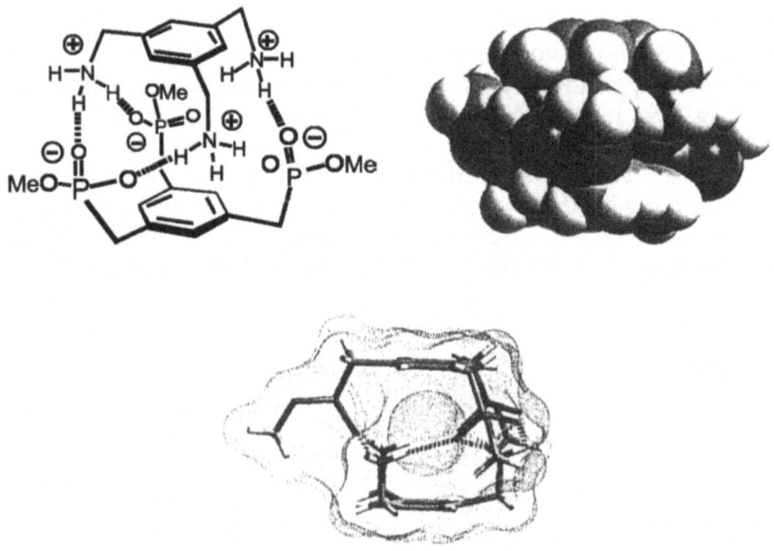

Fig. 41 Energy-minimized structure of the complex between a trisphosphonate salt and its analogous triammonium salt showing (*from left to right*) the Lewis structure, a front view of the CPK model and the solvent-accessible area around the complex with an internal cavity. Reprinted with permission from [119]. Copyright 2002 American Chemical Society

sule. Analogous structures based on a calix[4]arene core have also been reported [120]. Although the formation of the capsular complex was confirmed, no guest encapsulation studies have been reported yet.

Another recent example of an electrostatically joined capsules by Reinhoudt and co-workers, shows that it is possible to form a capsule by the use of interactions between a positively charged porphyrin (incorporating *N*-alkylpyridinium groups) and a sulphonato-calix[4]arene. The calixarene sits as a cap on the porphyrin and a small solvent molecule can be trapped within the resulting cavity (Fig. 42) [121]. The formation of the capped porphyrin was followed by UV absorption measurements in methanol which gave an association constant of approximately 10^7 M^{-1} which remained largely unchanged when the R group of the calixarene was altered. Similar capsules, between tetraamidinium and tetrasulphonato calix[4]arenes, have association constants in the range of 10^6 M^{-1} in MeOH (Fig. 43). The inclusion of tetramethylammonium and N-methylquinuclidinium salts and acetylcholine was evidenced by ^{1}H-NMR and ESI-MS measurements [122].

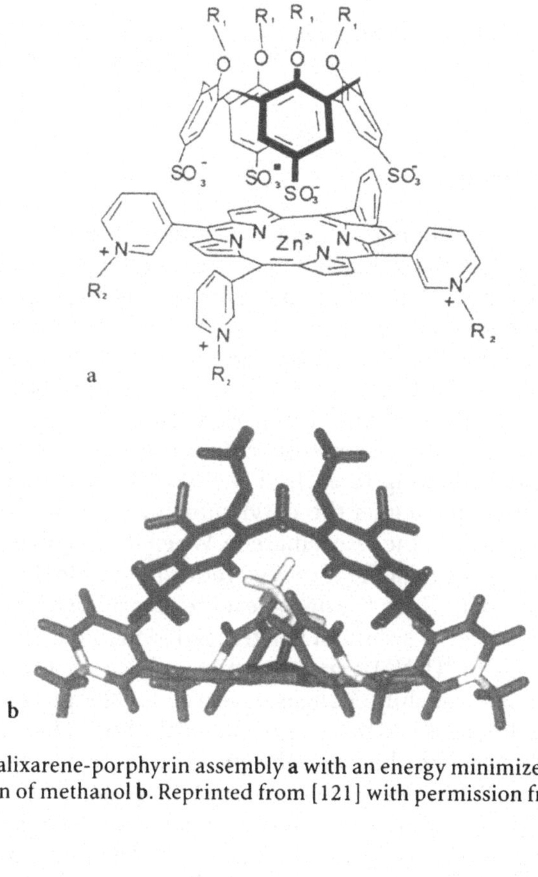

Fig. 42a, b The calixarene-porphyrin assembly **a** with an energy minimized structure showing the encapsulation of methanol **b**. Reprinted from [121] with permission from Elsevier Science

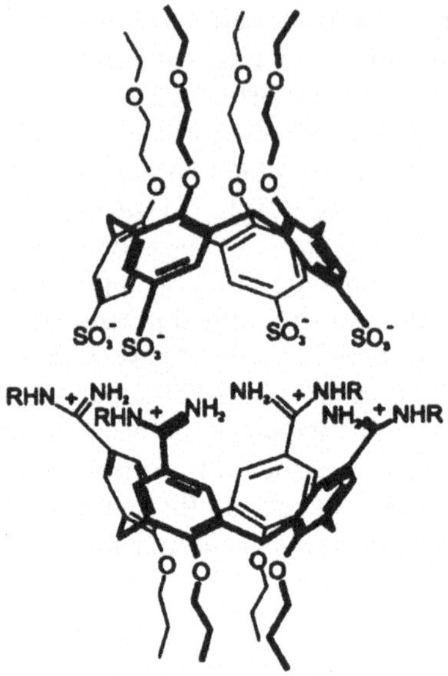

Fig. 43 A capsule formed between tetraamidinium and tetrasulphonato calix[4]arenes. Reprinted with permission from [122]. Copyright 2002 American Chemical Society

4.2
Self-Complimentary Capsules

Although assemblies containing a number of different species have been successfully formed, much of the research into non-covalent assemblies is concerned with the coming together of 'self-complimentary' species, those which can recognise and bind to themselves in solution [21, 24, 25, 123]. The first example of such a molecule that reversibly forms a capsule around its guest was a host for alkyl glucopyranosides [124]. The host monomer is a resorcinarene based system with eight alcohol functions situated on the upper rim (Fig. 44).

With an octyl chain on the glucopyranoside a host:guest ratio of 1:4 was observed, with one guest binding to each of the four diol binding sites situated around the resorcinarene ring as the octyl substituent makes the glucopyranoside too large to fit within the resorcinarene. When the length of the chain was reduced to a methyl substituent the observed ratio changed to 2:1. This ratio corresponds to a situation in which the one guest is encapsulated by the coming together of two resorcinarenes around it in a templated manner. The structure was established by the use of ^{1}H-NMR spectroscopy and vapour pressure osmometry. The ^{1}H-NMR chemical shift changes observed for the protons of the guest suggest a very close proximity to the aryl groups of the host. There are also strong NOE correlations between the host and the guest. The actual positions of the hydrogen atoms linking the two host molecules and the guest were not determined.

Hydroxyl residues have also been of use in assembling carcerand-type structures which are capable of reversible formation [125–127]. The monomeric species **53** can be reacted under different conditions to yield either the traditional carcerand

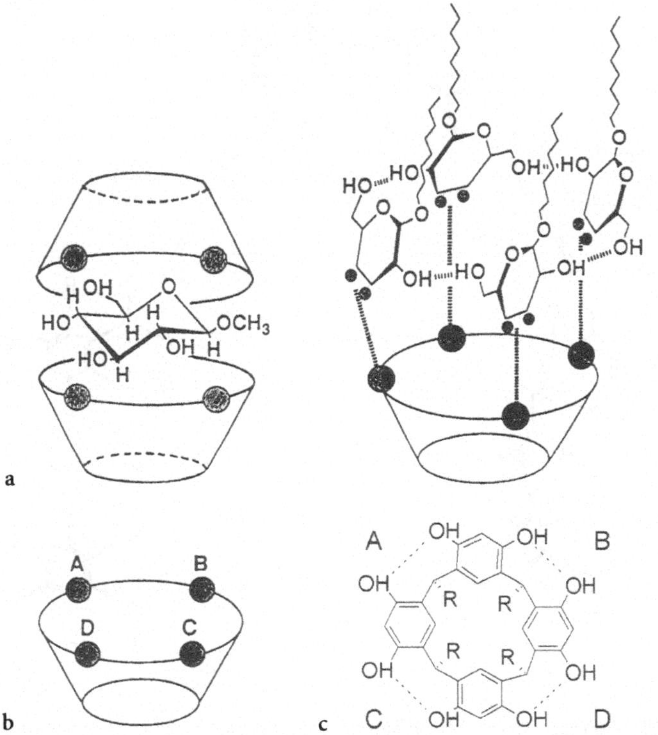

Fig. 44a, b The two ways in which the resorcinarene system can bind to alkyl glucopyranosides: **a** as a capsule with a 2:1 ratio; **b** as a monomeric host with a 1:4 ratio. Reprinted with permission from [124]. Copyright 1990 American Chemical Society

54 or the equivalent complex **55** that is held together solely by hydrogen bonds (Scheme 12). It was found that the guest species that acted as the most efficient templating reagents for the carceplex were also the most favourable guests for the hydrogen bonded capsule, with a preference shown for pyrazine. A crystal structure was obtained for the pyrazine complex of the hydrogen-bonded dimer (Fig. 45).

Rebek and co-workers have designed a cylindrical capsule based on the dimerisation of two expanded resorcinarene units which interact by eight bifurcated hydrogen bonds. The dimer has the ability to select one large molecule, such as terphenyl, or two small molecules, such as benzene and *p*-xylene or toluene, for guests (Fig. 46). Interestingly, the dimerisation of small self-complimentary pairs (such as 2-pyridone and benzoic acid) is also observed [128, 129]. These cylindrical capsules present several new features based around their encapsulation properties. Labile species, such as benzoyl peroxide, have been observed inside the capsule and can be stored inside for three days at a temperature of 70 °C, without decomposition and released when needed [130]. As well as stabilising reactive species, the cylindrical capsules can also act as reaction vessels, for example, the 1,3-dipolar cycloaddition between phenylazide and phenylacetylene which is observed to occur in a regioselective manner [131]. These

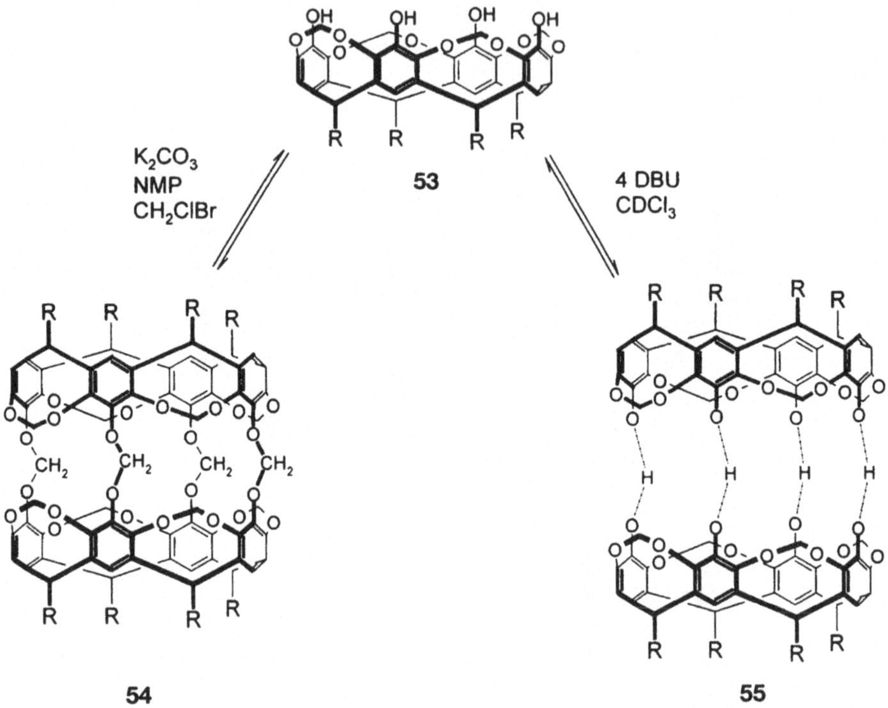

Scheme 12 Different reaction conditions can lead to either a carceplex **54** or a hydrogen bonded complex **55**

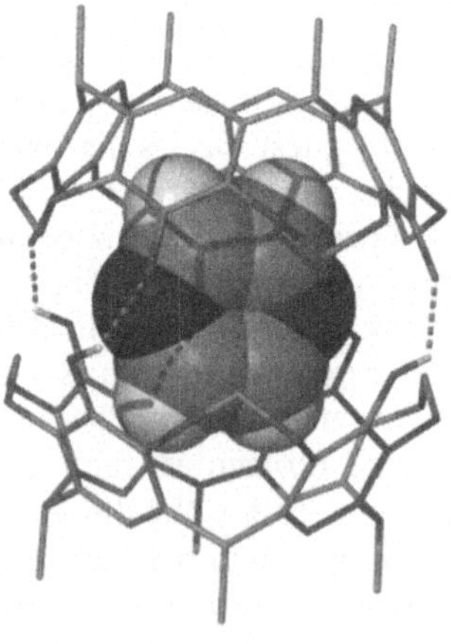

Fig. 45 X-ray crystal structure of the carcerand-like dimeric capsule 55 with pyrazine as a guest

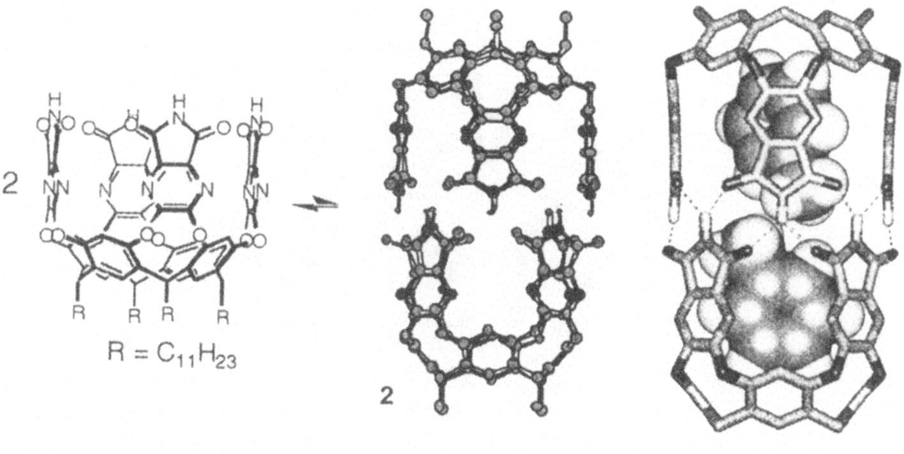

Fig. 46 The formation of a cylindrical capsule and its energy minimised structure containing two toluene guest molecules. Reprinted with permission from [131]. Copyright 2002 American Chemical Society

cylindrical capsules were also the first neutral cages to encapsulate anionic guest species, such as TsO^-, PF_6^- and IO_4^- [132].

Self-complimentary species are usually employed to form dimeric capsules, although not exclusively. Recently work has been carried out using a resorcinarene ring with eight hydroxyl residues around it, **56**, to form a hexameric, non-covalent assembly [133]. These hosts were tested using a range of tetra-alkylammonium (methyl-octyl) salts as potential guest species. The hexameric assembly, with a 6:1 host:guest ratio, was observed with the tetraheptyl, hexyl, pentyl, butyl and propyl salts, encapsulating both the tetraalkylammonium cation and the bromide counter anion. The formation of this species was determined by both ^{1}H-NMR and molecular modelling (Fig. 47). This is an example of the unpredictability of self-complimentary systems, there is often more than one possible aggregate that can be formed. A similar hexameric structure has been reported to be stable within a 50:50 mixture of water and acetone (Fig. 48) [111]. This robust assembly is very large consists of six molecules of **57** and has been observed by ^{1}H-NMR to en-

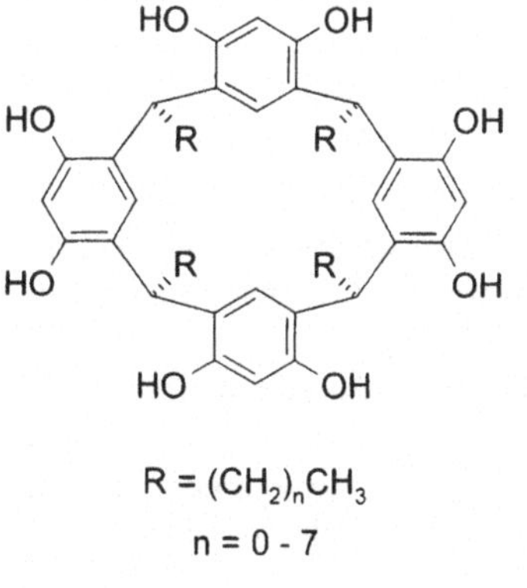

56

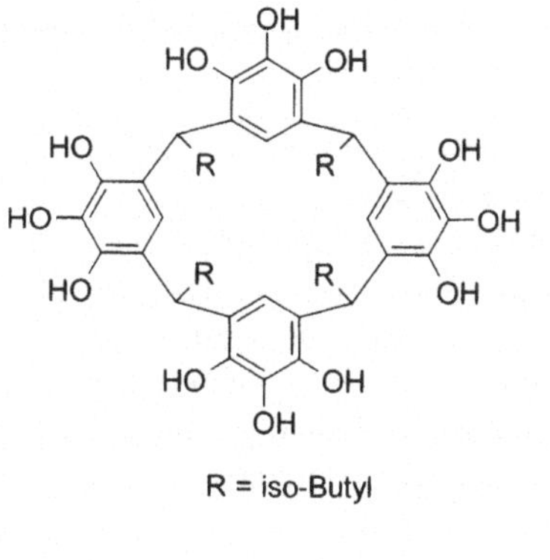

57

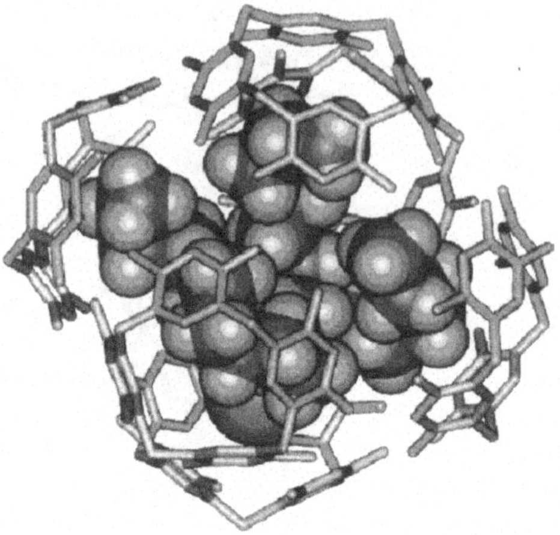

Fig. 47 Energy minimised representation of a resorcinarene hexamer of **56** with encapsulated $(C_7H_{15})_4N^+Br^-$. Reproduced with permission from [133]. Copyright 2001 National Academy of Sciences, USA

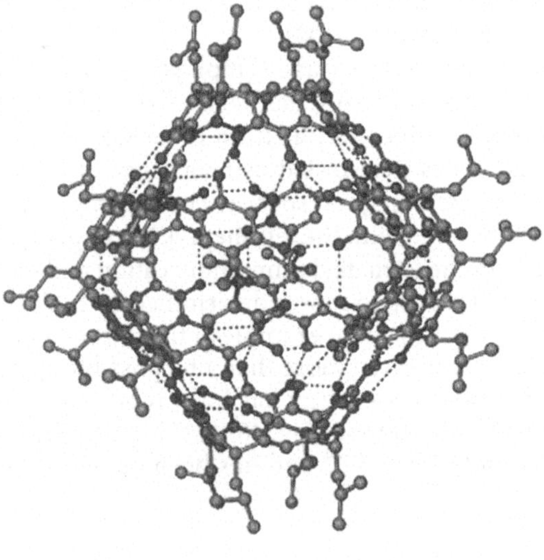

Fig. 48 X-ray crystal structure of the hexameric assembly of 57

capsulate 18 methanol molecules in solution. Other guest inclusion studies have been carried out on a similar system, suggesting that large assemblies of this kind can display exchange properties with a variety of guests [134]. Another hexameric structure has also recently been reported by Atwood [135].

The formation of capsules requires that the units to be joined together are of a suitable shape to form a cavity and that there are complementary groups through which hydrogen-bonding can take place. The easiest way to ensure complimentarity is to make one molecule that contains units which can both donate

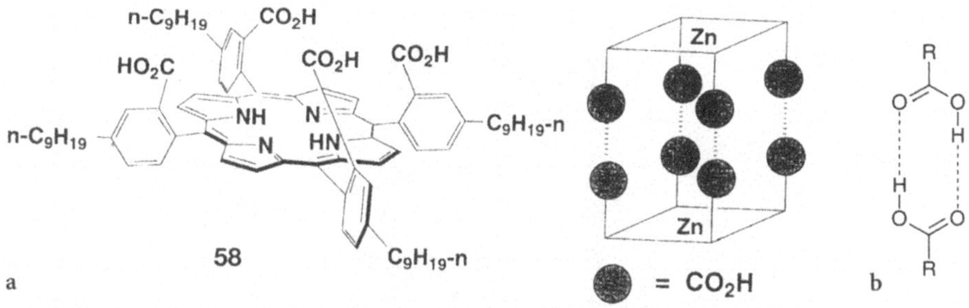

Fig. 49a, b A schematic representation of the porphyrin dimer $(\mathbf{58})_2$ held together by hydrogen bonds between the carboxylate groups, (reprinted with permission from [136], copyright 1997 American Chemical Society), and the way in which the carboxylate units join

and accept hydrogen bonds as shown in system $(\mathbf{58})_2$, which forms capsules between two carboxy-porphyrin derivatives [136]. A schematic representation of this capsule is shown in Fig. 49.

The capsule, as a di-zinc complex, was tested for its ability to bind pyrazine derivatives. These were used as guests because the two mutually *para* nitrogen atoms within the pyrazine ring have the potential to interact with the zinc contained in the porphyrin ring. The '4-arms up' αααα-atropoisomer of the porphyrin is exclusively involved in the formation of the dimer, and the amount of this conformer present, ascertainable by ^{1}H-NMR, is an indication of the existence of the dimer. ^{1}H-NMR and UV titrations were used to establish a binding constant for pyrazine within the capsule. The binding constant was found to be in excess of 10^7 M^{-1} in $CDCl_3$, making $(\mathbf{58})_2$ a very effective host. The size fit and the internal host-guest interactions provided by the Zn-N proximity are the major contributing factors to this binding strength. A surprising result was that even pyrazine derivatives with lengthy side chains also showed significant binding, with the side chain being able to protrude through one of the cavity walls (Fig. 50).

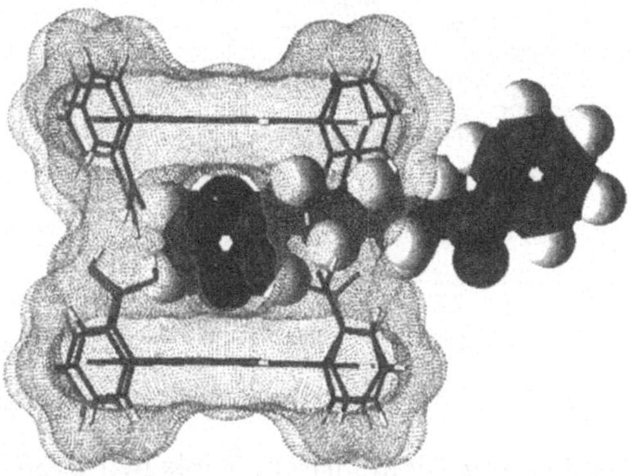

Fig. 50 A modelled simulation of the porphyrin dimer $(\mathbf{58})_2$ with a pyrazine derivative within it. The long side chain is able to remain outside the cavity. Reprinted with permission from [136]. Copyright 1997 American Chemical Society

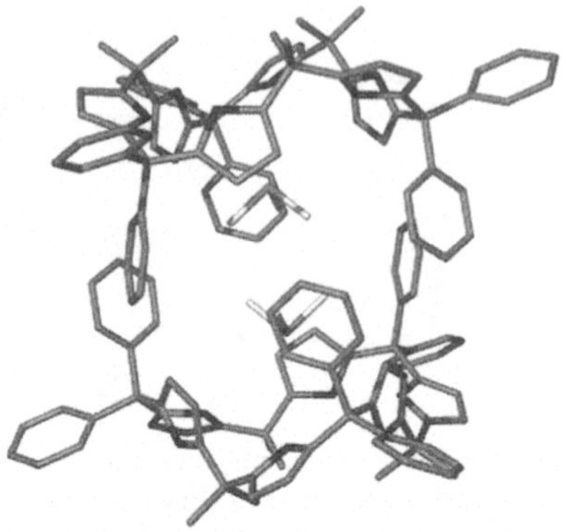

Fig. 51 Solid state structure of the $(59)_2$ dimer encapsulating two chloroform molecules

As with the multi-species components, it is not necessary for hydrogen bonding to be used in order to form dimeric capsules. This is highlighted by the recent synthesis of a molecule which assembles into dimers that are held together by π-interactions [137]. The *meso*-hexaphenyl calix[6]pyrrole **59** assembles into a dimeric capsule via the interactions between six phenyl groups arranged in a central belt around the capsule. The capsule is capable of holding two molecules of chloroform simultaneously, as seen in the solid state structure (Fig. 51). Similar phenyl interactions have been observed in a *p*-phenylcalix[5]arene which has been shown to dimerise and include C_{60} within its cavity (Fig. 52) [138].

Although there are many non-covalent systems already developed, and a great deal of scope for more, two systems have proved to be dominant in this area. These are systems containing glycoluril moieties (see next section) and those utilising urea as linking groups (see below).

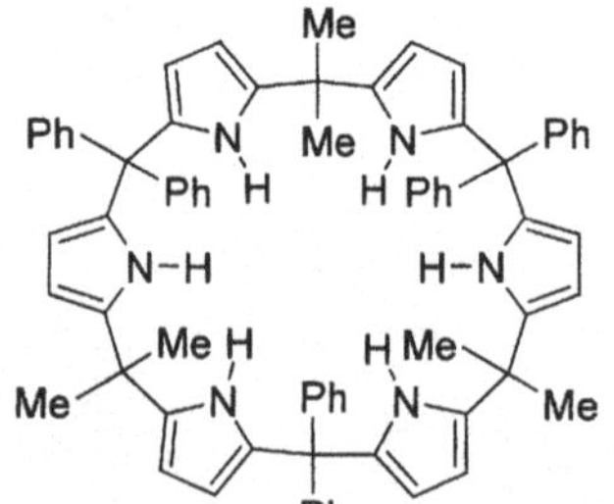

59

4.3
Glycoluril Systems

Glycoluril units have the capacity to form multiple hydrogen bonds by utilising the NH and C=O groups that they contain. Glycoluril units also have an intrinsic curvature to them, helping to form shapes that are mutually compatible (Fig. 53).

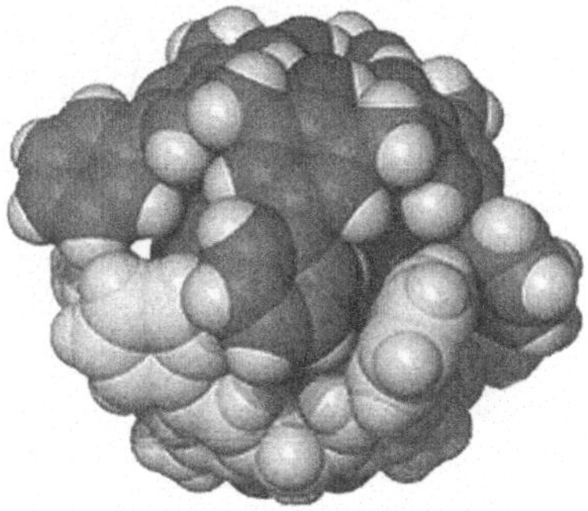

Fig. 52 The encapsulation of C_{60} by a *p*-phenylcalix[5]arene dimer [138], reproduced by permission of The Royal Society of Chemistry

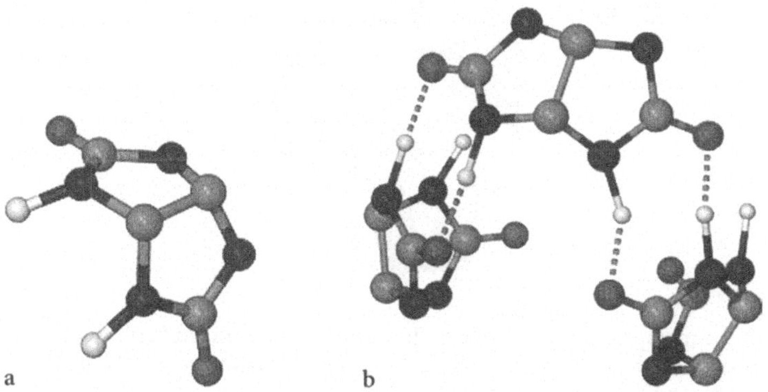

Fig. 53 **a** The glycoluril subunit. **b** The manner in which it can self-assemble

The first such system to be produced was the small curved molecule **60** containing two glycoluril units which, when assembled into a dimer, has been referred to as a 'tennis ball', so called because the hydrogen bonded 'seams' within the molecule can be imagined to be geometrically similar to those of a tennis ball (Fig. 54) [139]. This dimer was initially observed via ^{1}H-NMR and subsequently confirmed by X-ray crystallography. The structure of the tennis ball showed the presence of strong hydrogen bonds (NH···O, 167–178°, 2.78–2.89 Å). Guest species such as methane [140] and xenon [141] could be included within the dimers as demonstrated by upfield shifts in the spectra of the encapsulated guest. As with most systems that make use of hydrogen bonding, it proved essential that non-polar solvents are used in order to observe dimer formation. These solvents are less competitive for the hydrogen bonding sites, enabling composite species to form. In a few cases however, association has been seen in DMF [141]. The tennis ball, being the first system of its kind, is now the most understood and its guest exchange dynamics have been investigated by extensive use of two dimensional NMR [142], which shows strong correlations between groups which cannot be physically connected but which are brought into close proximity upon

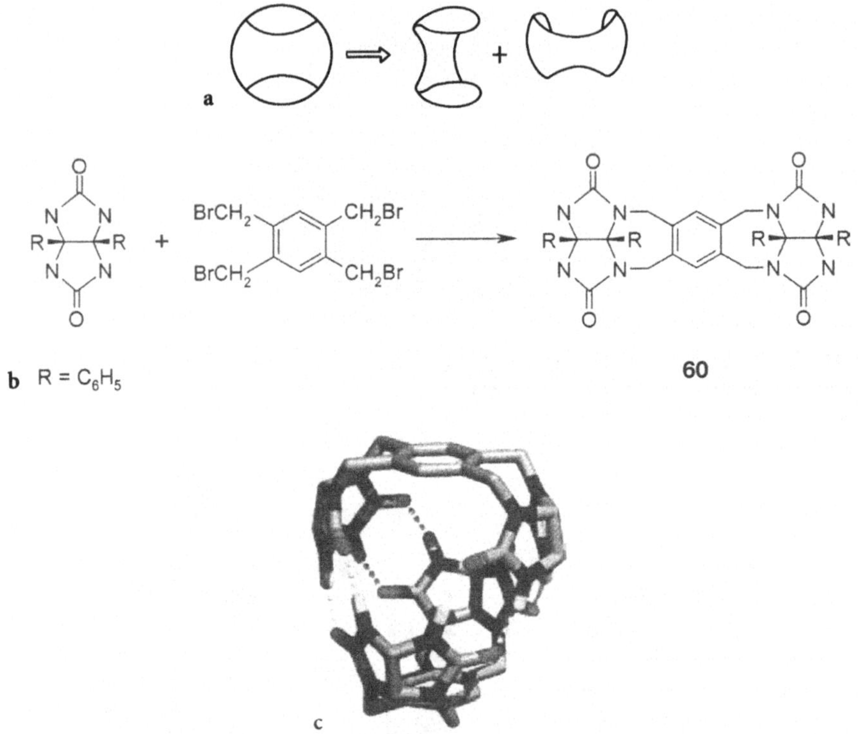

Fig. 54a–c The: **a** metaphorical synthesis; **b** chemical synthesis of the 'tennis ball' dimer; **c** the crystal structure with the phenyl groups omitted Reprinted with permission from [24]. Copyright 1999 American Chemical Society

dimerisation [143]. An analogously shaped inorganic system has also recently been developed [144].

The fundamental molecular framework of the tennis ball type of compound is very adaptable and can be readily altered to give a wide variety of analogues. For example, the electronic nature of the cavity has been modified by changing the substituent groups on the spacer between the two glycoluril arms. If the spacer is changed to quinone or hydroquinone then different affinities for guest species are observed. The size of the cavity is also subject to alteration via the nature of the spacer group which was changed from the original C_6H_2 spacer to those displayed in Fig. 55 [145]. The alkene-linked monomer **62**, the shortest, showed dimerisation behaviour but was unable to bind guests strongly, even the small guest methane, due to the decreased size of the cavity. The longer spacer units result in indeterminate behaviour due to the high degree of lability of the resulting dimers. These systems all showed evidence of heterodimerisation with **60** to produce irregularly shaped capsules. Evidence for these dimers was from ^{1}H-NMR signals assigned to encapsulated solvent, rather than a clear set of signals for the dimeric species itself.

Following the success of the tennis ball and its derivatives as effective dimeric capsules, the spacer group was further expanded so that a capsule could be formed

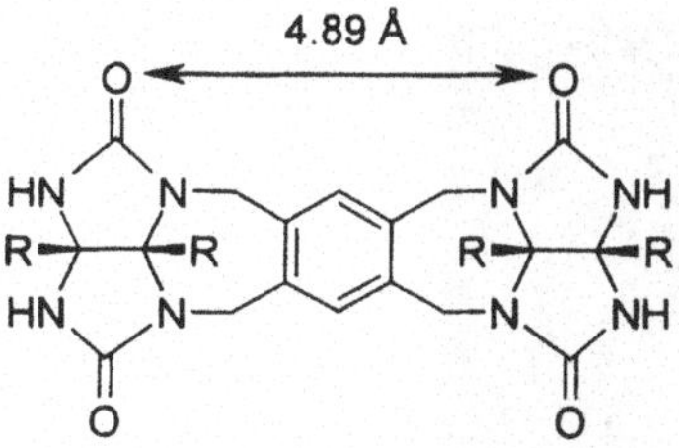

60

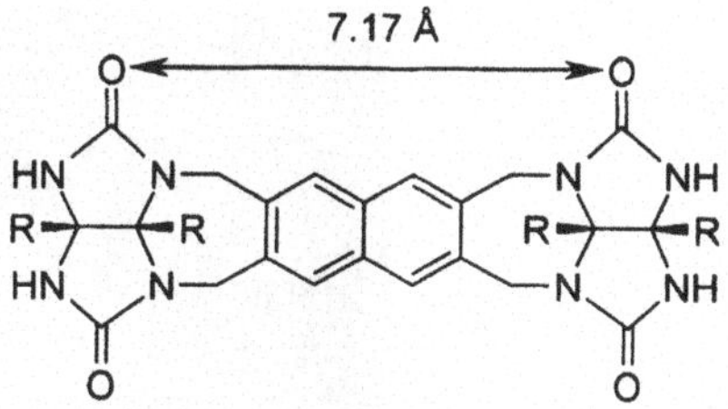

61

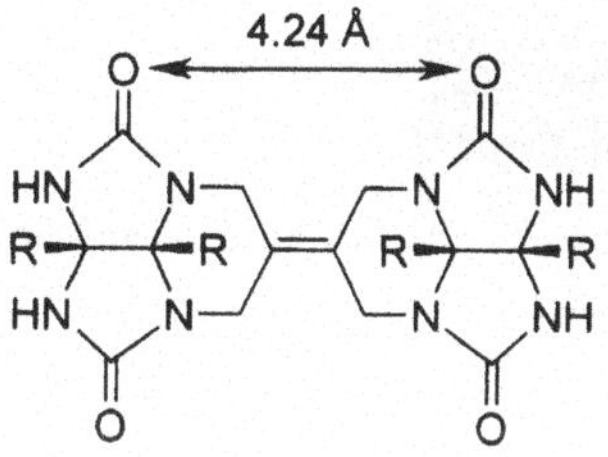

62

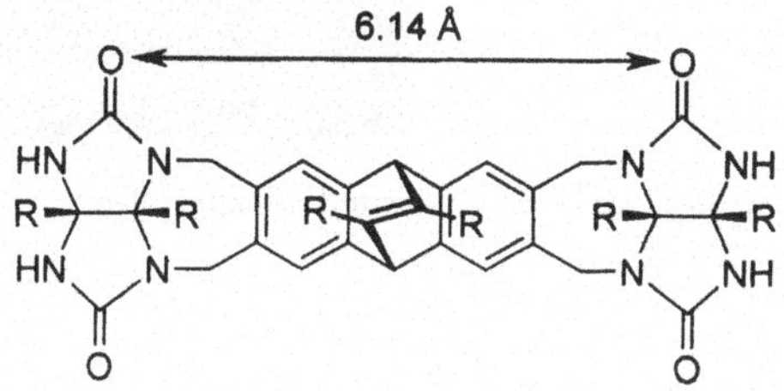

63

Fig. 55 Different spacers used to test heterodimerisation, **60** is the original tennis ball monomer

with an appropriate size to be able to include larger species than methane. The result of this approach was the formation of the 'softball' dimer from the monomer **64** [146, 147]. ^{1}H-NMR experiments showed that the best guests for this dimeric species are ferrocenecarboxylic acid and 1-adamantanecarboxylic acid. The process of encapsulation for the softball proved to be entropically driven. In order for a guest to be bound within the capsule, two molecules of solvent must first be ejected. Evidence came from the use of a two-solvent mixture in which three independent capsule species were observed with either two molecules of the same solvent or a mixture inside of them. There has also been a suggestion about the method by which guest exchange occurs, supported by molecular mechanics calculations [148]. This mechanism suggests that the dimer maintains some intermolecular hydrogen-bonds and opens up in a 'hinged' process to allow the exit of one guest followed by the entry of the replacement.

The addition of phenol groups along the 'side' of the monomeric species was found to increase the self-association of **64** by a factor of 10, by creating an additional 8 hydrogen bonds between the halves of the dimeric species. The stable

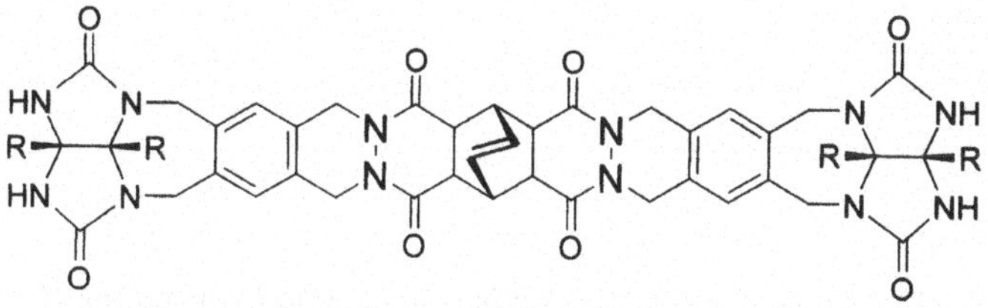

64

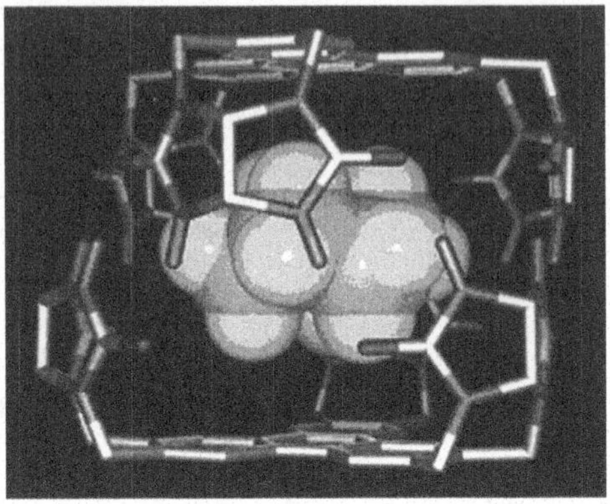

Fig. 56 The 'jelly doughnut' encapsulating cyclohexane, reproduced with kind permission from [153]

softball dimer has also been used as a molecular reaction chamber. A Diels-Alder reaction was observed to take place between *p*-quinone and cyclohexadiene, within the capsule [149–151].

The success of glycoluril units within these 'ball' systems has prompted research into related species. Keeping with the use of quirky names for molecules, one analogue is the 'jelly doughnut' [152]. Unlike the 'balls' which were both constructed from topologically linear components, the doughnut is based upon a monomeric unit that contains three glycolurils, equally spaced around a central triphenylene group. The shape of this new dimer is a flattened cylinder, somewhat like that of a doughnut, held together by twelve strong hydrogen bonds. The cavity within this capsule is of an appropriate shape to accommodate flat, circular guests such as benzene and cyclohexane (Fig. 56), and the host has shown selectivity for these guest species [153]. The exchange of these guests occurs on a timescale of several hours. Interestingly, the encapsulated cyclohexane exhibits slow ring-inversion as a result of the restricted space within which it is held [154].

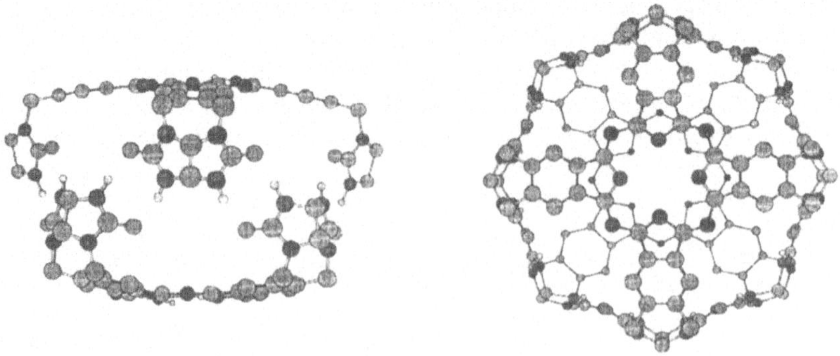

Fig. 57 Proposed structure of the phthalocyanine dimer $(\mathbf{65})_2$, reprinted from [155] with permission from Elsevier Science

The doughnut geometry of the host was taken a step further by use of a functionally active spacer group. Instead of simple aromatic spacers previously used, phthalocyanines were employed [155]. The geometry of the resulting dimer **(65)**$_2$ is similar to that of the doughnut, and dimerisation is observed to be preferable in aromatic solvents which will fit the cavity and template the dimerisation process (Fig. 57). The use of the phthalocyanine introduces a potential catalytic site within the capsule. To date, no work has been published investigating the catalytic potential of this dimer.

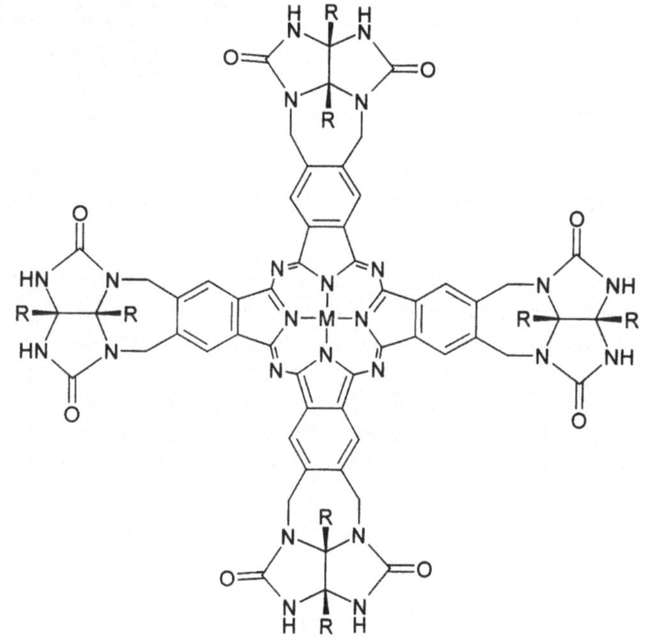

65

One of the most interesting recent developments with glycoluril capsules is the synthesis of chiral softballs (Fig. 58) [156]. When synthesising the monomeric units, one side of the spacer was made longer than the other side. Although these monomers are achiral, when dimerisation occurs, the result is a chiral aggregate with an asymmetric cavity. The guests tested with the chiral capsules were mainly camphor derivatives. The complexation diastereomeric excesses observed with such guests did not exceed 60%. This represents relatively poor selectivity for any useful system but remains an important step towards chiral recognition.

The majority of work investigating self-assembling capsules has been centred around reasonably rigid monomers which are conformationally restricted. The softball, **(64)**$_2$, whilst having freedom to fold over, cannot twist or distort due to the rigid ring systems. There are advantages, however, to using a more flexible system as this may allow the shape of the capsule to alter slightly to fit around the guest. Towards this end the glycoluril unit was appended in a flexible way to a triethylbenzene core [157], a spacer also frequently employed by other groups [158,

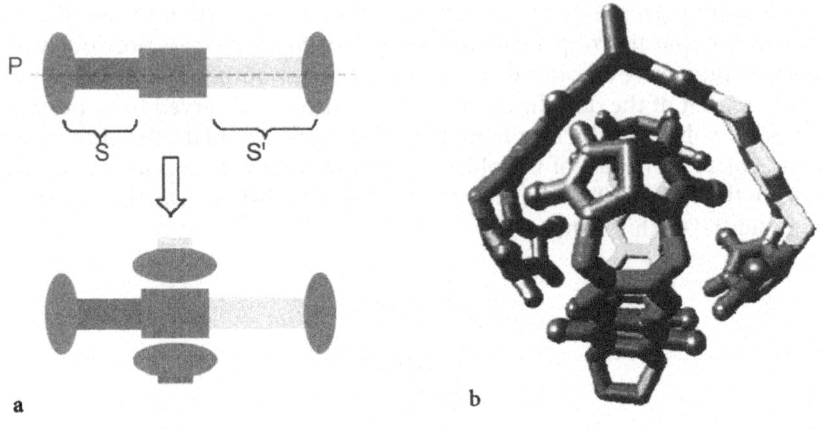

Fig. 58a, b By using two different spacers (S and S′) the resulting dimeric species is chiral. Reprinted with permission from [156]. Copyright 2001 American Chemical Society

159]. Many different systems of this general type have been prepared, using various different functionalities within the arms themselves (**66–69**). All of these species, except for **67**, were found to dimerise and to exist exclusively as dimers in non-competitive solvents. Mixtures of the monomers resulted in heterodimers being observed in most cases. The cavities have a volume of about 0.5 nm^3 and there are holes that are of sufficient size to allow the interchange of guests (Fig. 59). Larger guests form more kinetically stable complexes as they can

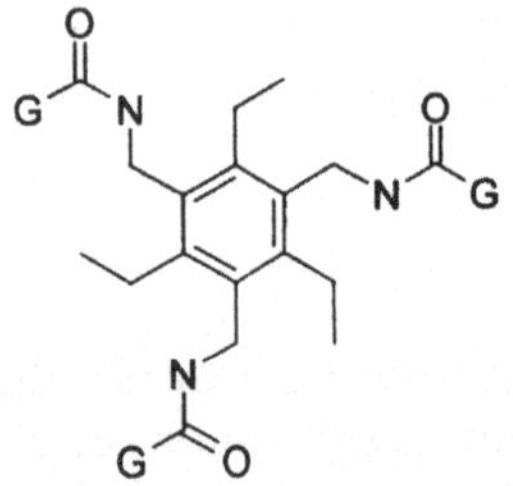

66

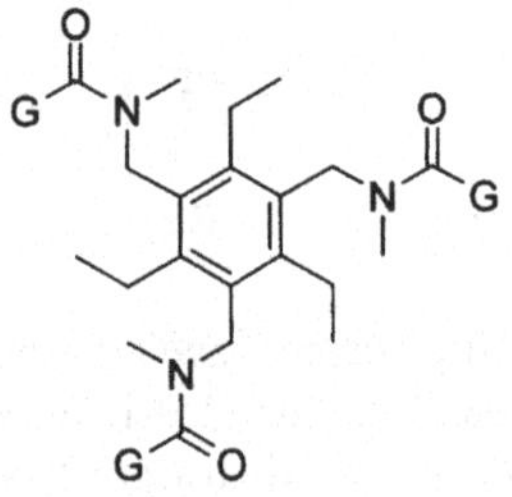

67

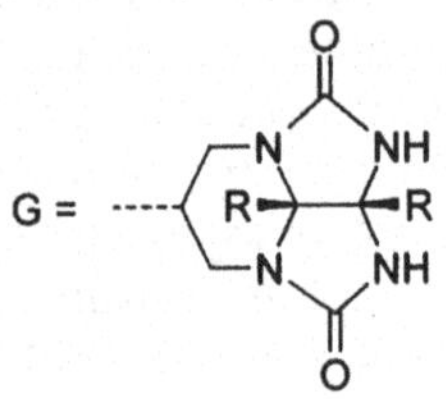

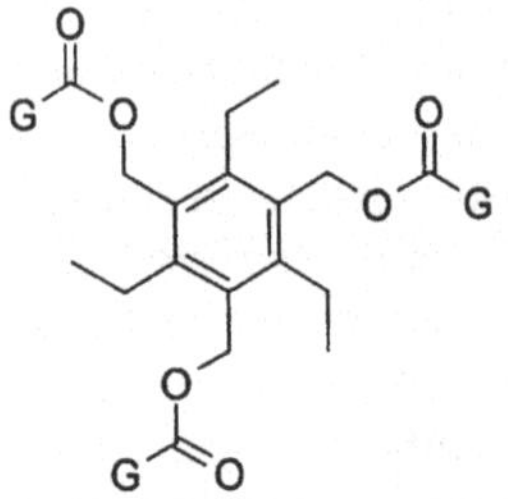

68

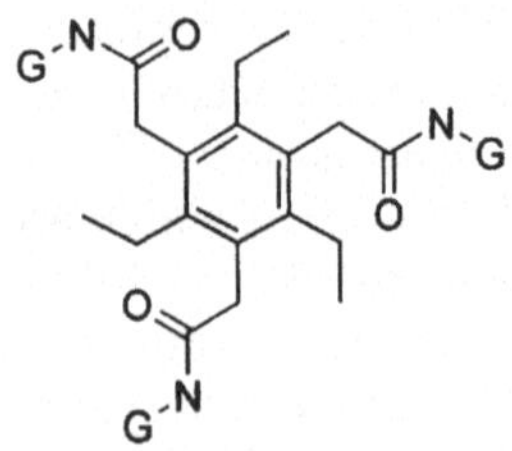

69

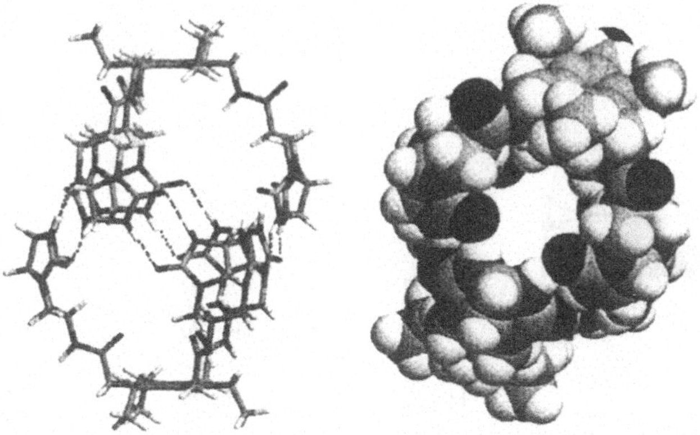

Fig. 59 Computed structures of a heterodimer between two flexible species (**66**) (**69**) with side chains omitted for clarity. Reprinted with permission from [157]. Copyright 2001 American Chemical Society

only enter and exit the cavity when the two halves of the dimer separate, a process which is slow on the ^{1}H-NMR timescale.

4.4
Urea Containing Capsules

The urea functionality also shows a strong tendency towards self-association, as in the helical channel structure of urea clathrates [160] and the cocrystallisation of diarylureas [161]. Initial studies using urea-derivatised frameworks were focused on the binding of anions within neutral receptors. Reinhoudt et al. produced the first such work based on calix[4]arenes derivatised on the lower rim [162]. This study was concerned primarily with the binding of halides. A later study was based upon a tri-derivatised calix[6]arene that is capable of binding larger anions, such as tricarboxylic acid salts [163]. Eventually this work led to the testing of these compounds as agents for transporting anions across membranes [164]. Since these initial displays of the usefulness of urea groups within synthetic host systems a large number of hydrogen-bonded capsules have been prepared and studied that contain urea functionalities as agents to join the halves of the capsule together. These compounds have mostly been based upon either calixarene or resorcinarene frameworks to provide a rigid backbone.

The first examples of urea-containing dimeric capsules were discovered independently by Rebek and Böhmer [165, 166]. These are based on calix[4]arene ethers, in which the lower rim ether units help to fix the calixarenes in a cone conformation via intramolecular interactions. Figure 60 shows the generic structure of such systems. Synthetically, the urea calixarenes are readily prepared. The calixarenes are first nitrated, followed by a reduction of these groups into amines. The urea derivatives are fixed to the upper rim by reaction of the *p*-amino calixarenes with isocyanates. Many systems were studied using differently substituted ureas which contain either short alkyl chains or simple phenyl derivatives. Studies also involved the changing of the lower rim ether substituents.

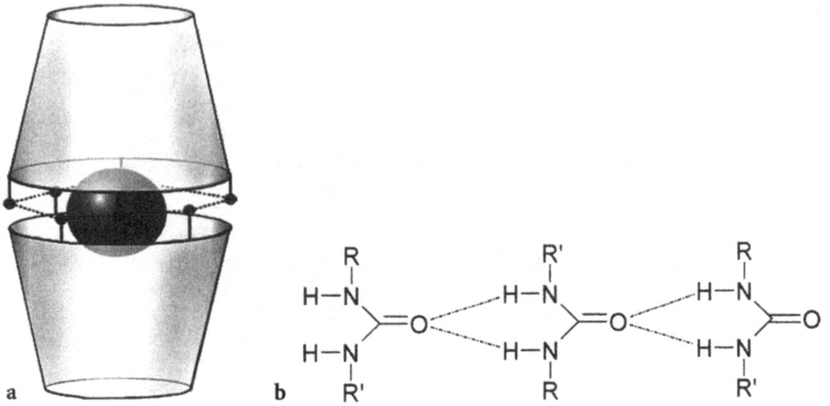

Fig. 60a, b A generic calixarene dimer and a schematic picture of the manner in which the two monomers form a joining seam

Evidence for dimerisation in these systems initially came from the unusual appearance of the ^{1}H-NMR spectra. The spectra recorded in DMSO-d_6 correspond to the free monomer, whereas the spectra run in $CDCl_3$ showed extra signals. The C_4 symmetry displayed in the $CDCl_3$ ^{1}H-NMR rules out the possibility of a pinched cone conformation with intramolecular hydrogen bonding. This left only the possibility of the calixarenes bonding *inter*molecularly, i.e. forming a capsule. The difference in behaviour between the two solvents used is explained by the fact that DMSO is a much more competitive solvent for interactions such as hydrogen bonds and therefore prevents dimers from forming. The work by the Rebek group was also backed up by their ability to show the encapsulation of a guest species within the dimer by ^{1}H-NMR. This is demonstrated by a peak appearing slightly downfield of the residual solvent peak in the spectra which was found to grow when protonated solvent was added.

Whilst Rebek used the inclusion of a guest as secondary proof of the capsules, Böhmer used the formation of heterodimers to show that the NMR spectra seen were not solely due to the monomeric species. In DMSO-d_6, only a mixture of the two species was seen. However, the use of a non-polar solvent showed sets of peaks corresponding to all the possible dimers (AA, BB and AB) in the expected statistical mix.

Around the same time as this, Reinhoudt developed a calix[4]arene system with only two urea or thiourea functionalities attached on opposite faces **70–72** [167]. These less substituted systems display both *inter*- and *intra*molecular hydrogen bonding as a result of the calixarene adopting a pinched cone conformation (demonstrated by the use of NOESY NMR). In some spectra it is impossible for the connectivities to be made within a single molecule, so the only possibility left is that dimerisation occurs. As with the initial experiments of Rebek and Böhmer, the extent of hydrogen bonding was observed to be solvent dependent. Concentration dependant FTIR was also used, to observe the effects on the NH stretching vibrations, but no concentration dependence was observed. Of the three urea derivatives used, only **72** showed no evidence of dimerisation

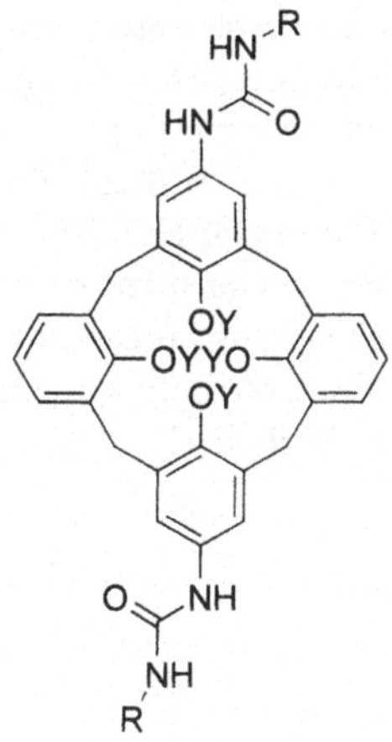

70 R = phenyl Y = CH_2CO_2Et
71 R = *t*-butyl Y = CH_2CO_2Et
72 R = *n*-octyl Y = CH_2CO_2Et

with either the urea or thiourea moiety. No reason was suggested for this although it is likely that there is too much conformational freedom in the group to allow a highly organised structure to form.

Unfortunately, as with many systems of this type, crystal structures are not readily available. This means that molecular modelling must be employed to try and fully understand the systems. An energy minimised structure of 72 in a monomeric state does indeed show intramolecular hydrogen bonding between the urea groups (Fig. 61). No structure was presented for the dimer.

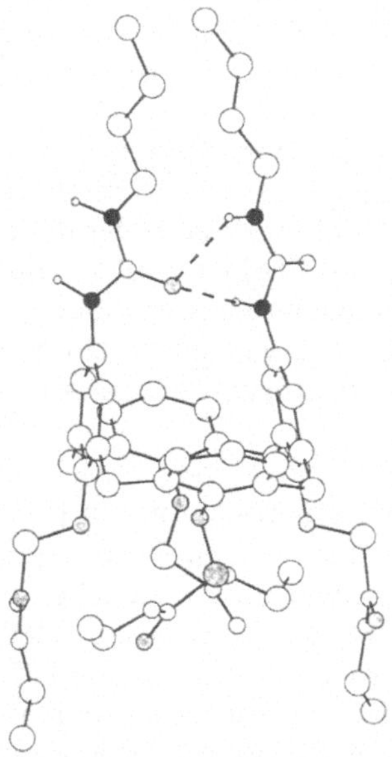

Fig. 61 Energy minimised structure of the two armed calix[4]arene species 72. The long alkyl chains are only partially shown. Reprinted with permission from [167]. Copyright 1996 American Chemical Society

These first examples led to a mass of work by these contributors in the following years. Böhmer has carried out NMR studies on reversible guest inclusion within calixarene dimers [168]. By using different ether groups around the lower rings, the symmetry of the calixarenes was reduced to make the spectra less ambiguous, for example **73**. This also allowed the use of NOESY experiments to determine exchange rates of guests and the rate constant for dimerisation. The dimerisation rate of **73** was found to be 0.26 (6) s^{-1} with the exchange of a benzene guest occurring at 0.47 (10) s^{-1}. These results were confirmed by using a variety of NMR experiments (TOCSY, EXSY, ROESY, HMBC).

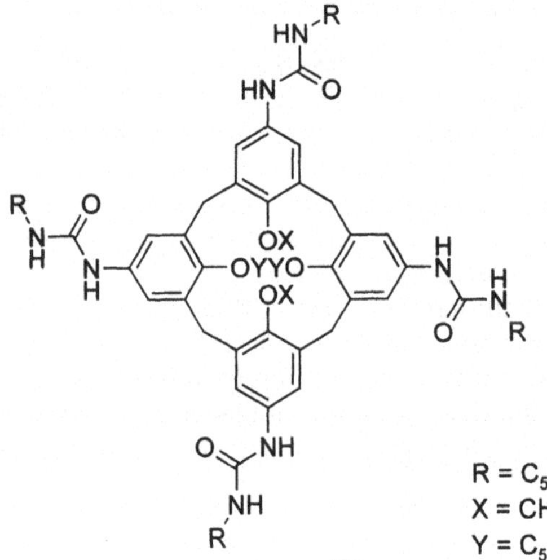

$R = C_5H_6$
$X = CH_3$
$Y = C_5H_{11}$

73

Reinhoudt has shown that it is possible to gather accurate information on the presence of dimers by utilising mass spectrometry [169]. Usually, weakly coordinated molecules will not hold together under the conditions imposed during MS analysis, even the very soft MALDI-TOF technique. The new method employed was to label the components using Ag^+ ions, which have a high affinity for aromatic π-donors. Although many results were obtained that showed the presence of a number of complex hydrogen-bonded assemblies, none were of dimeric species.

The kinetic stability of urea-calixarene dimers has been observed to be highly dependant on both the urea substituent group [170] and on the guest that is being included within the capsule [171]. The calix[4]arene system **75** was studied using 5 different substituent groups attached to the urea functionalities (Scheme 13). The monomeric species were all tested for signs of dimerisation both individually and with each other. Compound **75b** proved the least prone to aggregation and no dimers were observed. Compound **75e** with the triphenylmethyl substituent only formed dimers in 50% of cases tried and, interestingly, formed no homodimers.

Experiments have been carried out to study the kinetics of guest exchange within capsules of the type 75. The general procedure is for two monomers to be

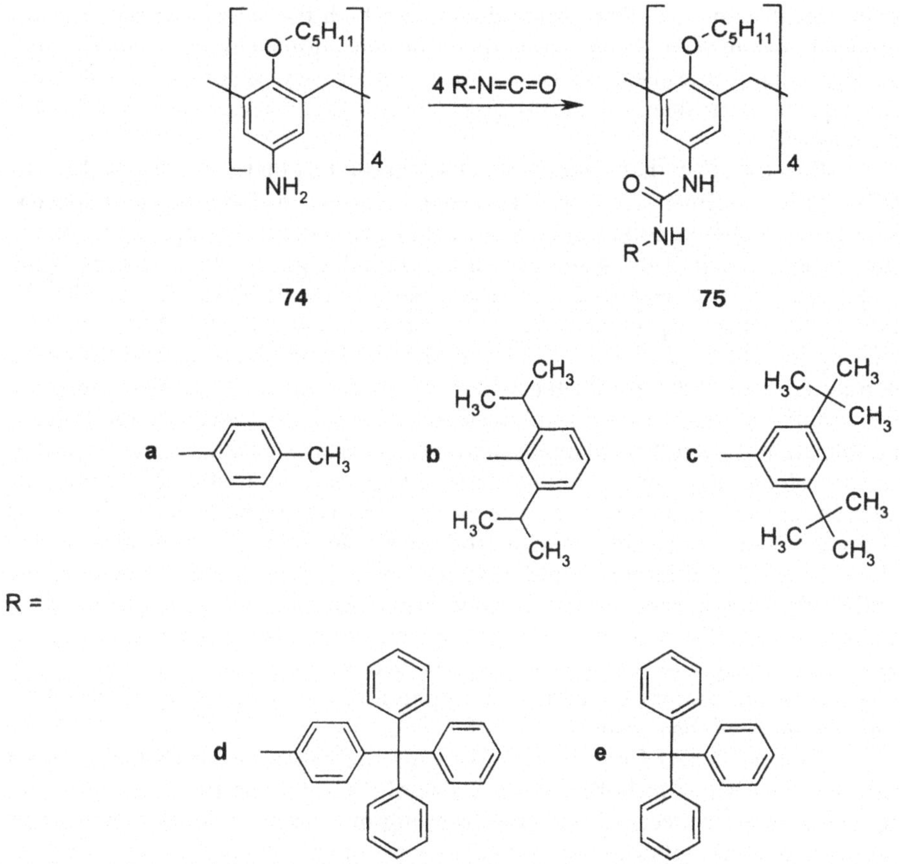

Scheme 13 The synthesis of differently substituted calixarene-urea capsules used for stability studies

dissolved in benzene, which was then allowed to evaporate to leave behind a solid. This residue was then dissolved in benzene-d_6, and was observed by ^{1}H-NMR over a period of time (~10 days). The spectra clearly showed a peak for the included benzene which gradually disappeared with first-order kinetics. Using systems **75d/e** and **75a/e**, half lives of 60 hours and 130 minutes respectively have been observed. This is a significant contrast, and can be attributed to the only difference between the two systems, the relative bulkiness of the substituent on the urea groups of one of the monomeric species. The experiments were also carried out using other deuterated solvents with similar structures, such as toluene and xylene. The behaviour observed for **75d/e** was very different in the presence of toluene, which does not form stable complexes with the heterodimers. Instead, the benzene heterodimers split apart to form the toluene homodimer $(\mathbf{75d})_2$. The most remarkable result found in this study is the kinetic stability of heterodimer **75d/e** in the presence of cyclohexane-d_{12}. The benzene containing dimer remains in a 1:1 guest:capsule ratio after 18 h at 40 °C. This can be compared to a control experiment using the $(\mathbf{75a})_2$ homodimer, in which the signal attributed to included benzene disappeared under the same conditions. This opens up the possibility of indefinite guest inclusion until a suitable agent is added that will cause the capsule to break apart (in this case a solvent that disrupts hydrogen bonds, such as DMSO).

To show the effect that a guest could have upon the kinetic stability of the capsule, the homodimer $(\mathbf{75a})_2$ was used with a variety of different guest species. The guest-included compounds were again produced via evaporation of the monomeric species from a solution of the intended guest. The exchange of the guest against cyclohexane-d_{12} was then followed by ^{1}H-NMR. The results obtained show that the presence of different guests can vary the half-life of the capsule significantly. A half-life of 2.9 h was observed with chloroform as the guest, compared with 1120 h for the tetrachloromethane capsule. These two guests are very similar in many respects and the factors that lead to such a difference in the lability are not yet fully explained, although it is suggested that the acidic proton in chloroform may interact with the urea functionalities to destabilise the capsule. Other work also showed marked differences in the stability of closely related compounds or inconsistencies with expected results. Whilst benzene and toluene show significant differences in their stabilities with capsule $(\mathbf{75a})_2$, cyclohexane and methylcyclohexane are remarkably similar. Another interesting trend is the increase in stability in going from benzene to fluorobenzene to 1,4-difluorobenzene. The strength of the hydrogen bonds formed (based on the chemical shift changes of the protons concerned) is apparently unrelated to the observed kinetic stability of the capsules.

So far the scope of guests examined for the calixarene systems has been limited. This is due to both the small size of the cavity and because the main focus has been concerned with the trapping of solvent molecules. While this work is interesting it does not immediately lead to many practical applications. Recently, work has been carried out that has led to larger capsules such as **76** (with an internal Van der Waals volume ca. 400 Å^3) being formed by extending the arms of the calixarenes [172]. Molecular modelling has shown that these hosts have the potential to include two benzene-sized guests. Extended capsule **76** also possesses a different cavity shape with the longer sides forcing a more cylindrical geometry, although not too far removed from the equivalent non-extended unit since heterodimers have been observed. Extended hosts have been made using *p*-tolyl and *p*-(*n*-hexyl)phenyl ureas, with the hexyl providing much better solubility. These capsules are able to form with bound sodium salts. The sodium ion is located at the lower rim of the calixarene groups close to the ether oxygen atoms. The exchange of the included solvent is very rapid. This fast exchange has been attributed to the fact that there are larger holes in the 'sides' of the capsule through which small solvent molecules can readily pass. This means that the formation of the capsule does not have a great effect upon the binding of guests and behaves in a similar way to hemicarcerands (see above).

In addition to being able to bind sodium salts, the binding of alkyl-pyridinium cations by **76** has also been observed. This class of guest exhibits binding constants of between 5×10^3 and 2×10^5 M^{-1} and displays fast exchange on the NMR

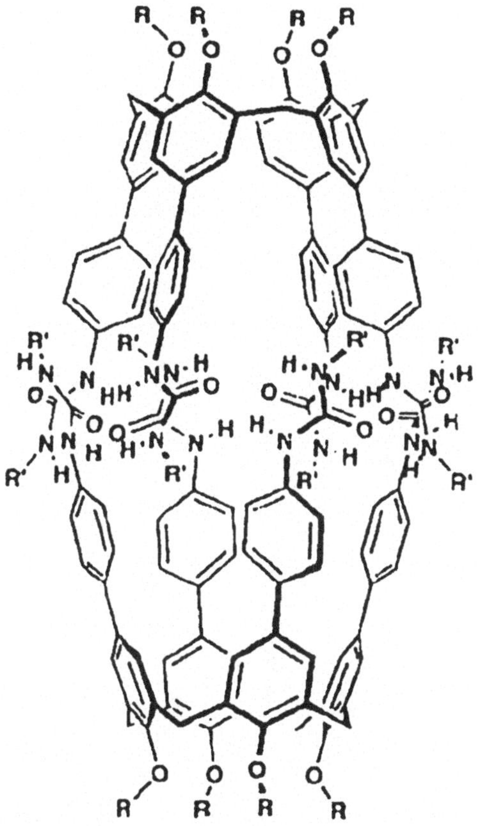

76

timescale. The use of the larger capsules as reaction chambers for catalysis has been suggested, although the fast exchange that is observed could limit the applicability of this approach.

A recent study [173] has shown that, in at least one case, the species trapped within a capsule is not frozen in position and its motion has been studied along with a remarkable change in the orientation of the hydrogen-bonded urea belt (Fig. 62). The ring of hydrogen bonds that runs around the equator of capsule $(77)_2$ is observed to switch direction, contrary to the previously observed static belts which are fixed in a mixture of the two orientations. This exchange is fast on the NMR timescale at 25 °C even though the capsule is kinetically stable under these conditions.

Initially it was believed that this anomalous behaviour was linked to the rotation of a tetraethylammonium guest within the capsule which has been observed. This proved not to be the case, as was confirmed by the use of lower symmetry calixarene derivatives to reduce the ambiguity of the NMR spectra. Molecular dynamics calculations were employed to show that the size of the tetraethylammonium guest means that the capsule has to be significantly expanded in order for it to fit within the cavity. The result of this expansion is that the ureas are only

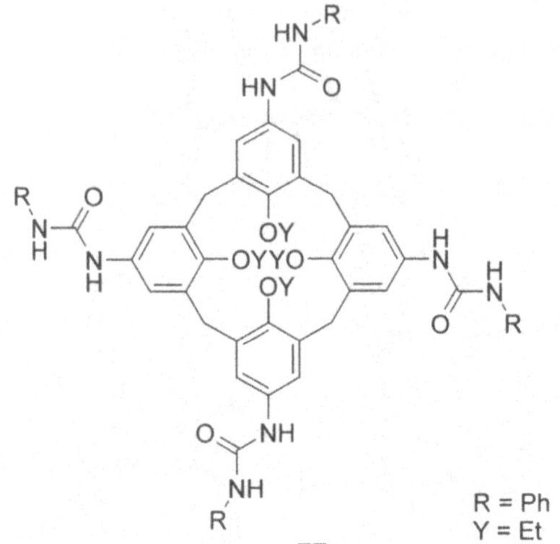

77

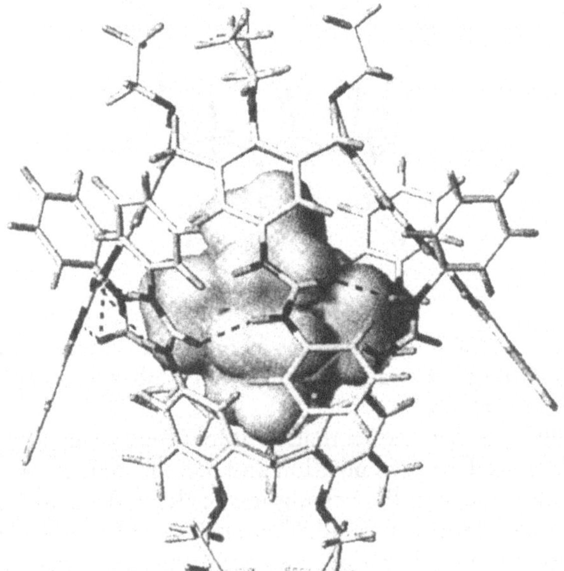

Fig. 62 Et_4N^+ encapsulated within the dimer $(77)_2$. This guest is observed to rotate within the host. Reproduced with permission from [173]

connected via one hydrogen bond instead of the usual two. The fact that the guest is cationic leads to the presence of cation-π interactions which are able to more than compensate for the reduction in affinity caused by the expanded geometry. This extra energy explains why the capsule can remain stable despite the fluxionality of the urea belt.

The vast majority of the work on urea calixarene systems has focused around calix[4]arenes, primarily due to their ability to adopt a rigid cone conformation, appropriate for the formation of a roughly hemispherical cavity. Calix[6]arenes are not as rigid, but this has not prevented researchers from trying to use them to form larger capsules. The calix[6]arene studies have used a system which contains three urea functionalities, appended to alternate phenolate rings [174]. They are more difficult to dimerise than their smaller counterparts, due to the lack of rigidity in the backbone and steric hindrance when joining the two halves together which prevents anything other than *N*-unsubstituted ureas from being used. The solution behaviour of these larger capsules is almost identical to the calix[4]arenes in terms of their behaviour in polar and non-polar solvents. The main difference, as expected, is in their guest encapsulating ability. The size of the cavity means that larger guests can be included.

Recently, urea based systems have been designed and synthesised by Alajarín et al. that deviate from the traditional rigid calixarene systems and are instead centred around a flexible tribenzylamine core [175, 176]. Compound **78** is a tris(*o*-ureabenzyl)amine which, in non-polar media, can self assemble into a hydrogen-bonded dimer (Fig. 63a) [175]. The R_1 groups examined have been *p*-tolyl and benzyl, of which the tolyl showed the greatest amount of dimer present in solution (>98% compared to 80–85%). ^{1}H-NMR and ESI-MS demonstrated the existence of the hydrogen-bonded species. With any more than 40% DMSO added to the solution no dimers were formed due to the competition from the solvent. In $CDCl_3$ the association constant of the two halves of the capsule was found to be in the region of 8.3×10^4 M^{-1} (R_1=*p*-tolyl). The capsule was found to be empty in solid state studies as it is too small for any molecules to fit inside. However, the substitution of the phenyl rings around the tertiary amine can be changed to give the tris(*m*-ureabenzyl)amine **79** [176].

Compound **79**, upon dimerisation, forms a larger capsule than that observed for **78** and is capable of encapsulating small guest species such as dichloromethane and chloroform (from a $C_2D_2Cl_4$ solution). The encapsulation was ob-

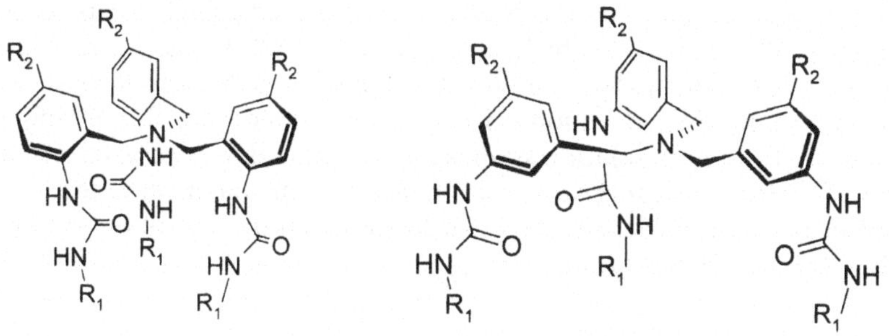

78 **79**

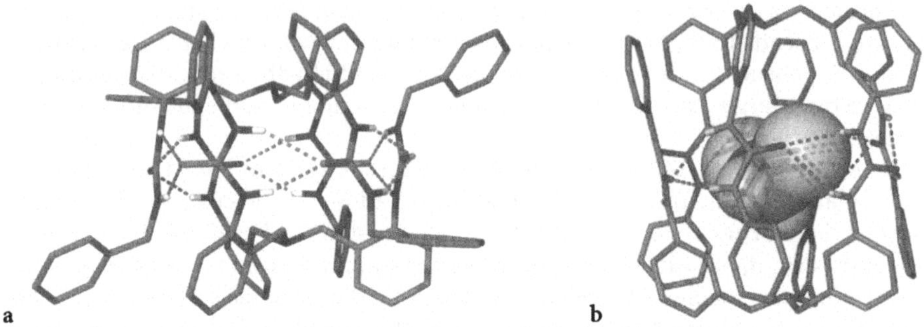

Fig. 63a, b X-ray structure of: **a** the empty $(\mathbf{78})_2$ dimer; **b** the dimer $(\mathbf{79})_2$ with encapsulated CH_2Cl_2 (right). Hydrogens not involved in H-bonding and R_2 groups are omitted for clarity

served by ^{1}H-NMR and confirmed by an X-ray structure (Fig. 63b). Benzene was not encapsulated from a chloroform solution but was in a $C_6D_6/C_2D_2Cl_4$ solution. Within the chloroform solution, the solvent molecules themselves are encapsulated with preference over benzene whereas with larger solvent molecules the benzene can enter. This highlights the way in which small substitutional alterations to a potential host species can significantly alter the size and shape of the internal cavity and hence influence the guest binding characteristics.

4.5 Unimolecular Capsules

Host molecules that completely surround other molecules make use of either strong covalent interactions, as in carcerands, or weak non-covalent interactions, as in self-assembling capsules. Unimolecular capsules blend the characteristics of the two. They are built by connecting two self-complimentary units with a flexible tether. In this process there is a loss of symmetry in the resulting dimer. Rebek has used this approach, connecting the two subunits of different dimeric systems such as softballs [177] and tetraureidocalix[4]arene-based capsules [178] to transform achiral molecules into chiral containers. The unimolecular capsule was referred to by Rebek as a *clam*. Although encapsulation with chiral guests such as pinanediol and camphorsultam led to a modest d.e. of 20%, this result bodes well for the use of the clam as a chiral catalytic reaction chamber.

Unimolecular capsules have also been prepared around a single resorcin[4]arene core, such as **80**, held together by hydrogen bonds [179]. These capsules form in the presence of tetramethylammonium chloride, whereby the cation exists within the cavity and the chloride anion exists outside. Evidence for this arrangement was obtained from both solution studies (^{1}H-NMR) and X-ray crystal structures (Fig. 64). Studies with either larger or smaller cationic partners to the chloride formed weaker, less stable complexes, indicating that the size of the cation may template the capsule formation.

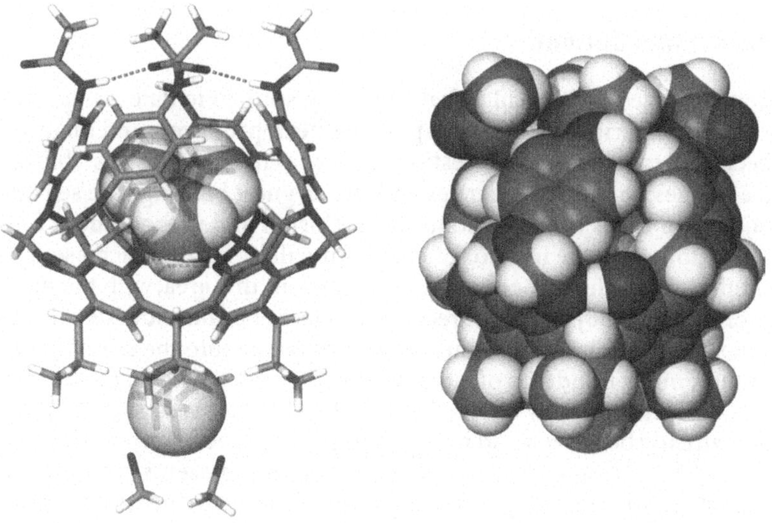

Fig. 64 Side view of the crystal structure of **80** binding tetramethylammonium chloride with the cation in the cavity and the anion located outside

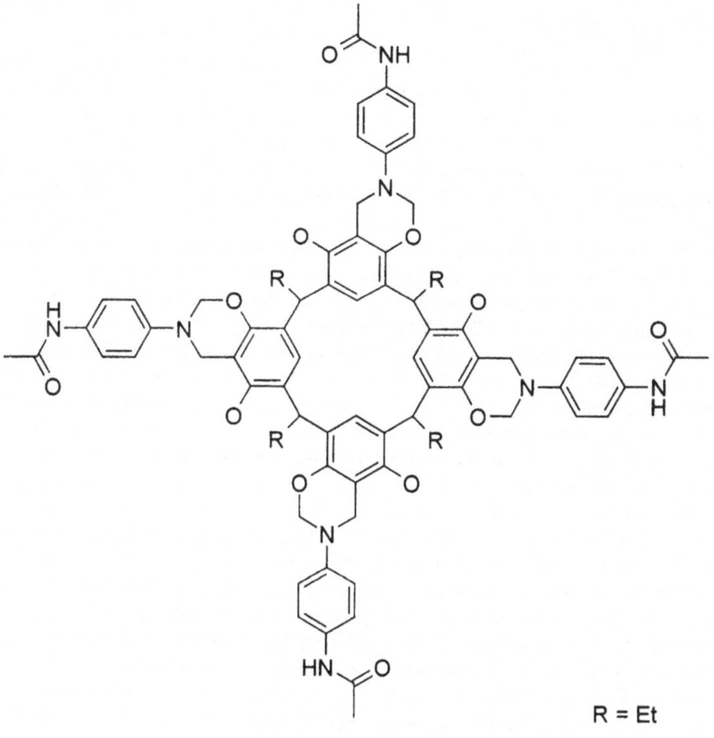

80

5
Conclusions and Outlook

The scope of cages and capsules that have been studied to date is massive. Cavities of widely differing sizes and shapes have been produced, and many of these have been shown to possess interesting properties in terms of the stabilisation of reactive species, catalytic properties and strong guest binding. The selectivity of the majority of cages is restricted to size exclusion and without more functionalities positioned on the interior of cavities it will be hard to increase the selectivity of these capsules. Progress has been made in this area, with the porphyrin system of Ogoshi [136] and more recently in a large hexameric system by Atwood which has hydrogen bonding hydroxyl groups facing into the cavity [180].

Although many hundreds of systems already exist there lies the potential for many thousands more, any of which could display the kinds of guest properties that have already been seen and probably many unexpected properties so far unthought of. The 3D encapsulation of guest species represents possibly the most discriminating approach to guest recognition if the cavity interiors can be accurately tailored for the potential guest species and the future looks bright for advances in guest inclusion studies.

Acknowledgements We would like to thank the EPSRC for funding (DRT) and Dr. Len Barbour for the program X-Seed. We also thank Dr. Agnieszka Szumna and Prof. Jerry Atwood for supplying Fig. 64. AP and MA acknowledge the support of the MYCT (Ramón y Cajal contract and project number BQU2001–0010).

6
References

1. Lehn J-M (1995) Supramolecular chemistry: concepts and perspectives. VCH, Weinheim
2. Beer PD, Gale PA, Smith DK (1999) Supramolecular chemistry. Oxford University Press, Oxford
3. Steed JW, Atwood JL (2000) Supramolecular chemistry. Wiley
4. Vogtle F (1991) Supramolecular chemistry. Wiley, Chichester
5. James TD, Hartly JH, Ward CJ (2000) J Chem Soc, Perkin Trans 1:3155
6. Sherman JC, Jasat A (1999) Chem Rev 99: 931
7. Cram DJ, Cram JM (1994) Container molecules and their guests. The Royal Society of Chemistry, Cambridge
8. Lindoy LF, Atkinson IM (2000) Self-assembly in supramolecular systems. The Royal Society of Chemistry, Cambridge
9. Lehn J-M (2002) Proc Nat Acad Sci USA 99:4763
10. Jones CJ (1998) Chem Soc Rev 27:289
11. Fujita M (1998) Chem Soc Rev 27:417
12. Olenyuk B, Fechtenkotter A, Stang PJ (1998) J Chem Soc, Dalton Trans 1707
13. Linton B, Hamilton AD (1997) Chem Rev 97:1669
14. Caulder DL, Raymond KN (1999) J Chem Soc Dalton Trans 1185
15. Davis AV, Yeh RM, Raymond KN (2002) Proc Nat Acad Sci USA 99:4793
16. Holliday BJ, Mirkin CA (2001) Angew Chem Int Ed 40:2022
17. MacGillivray LR, Atwood JL (1999) Angew Chem Int Ed 38:1018
18. Swiegers GF, Malefetse TJ (2000) Chem Rev 100:3483
19. Leininger S, Olenyuk B, Stang PJ (2000) Chem Rev 100:853
20. Swiegers GF, Malefetse TJ (2002) Coord Chem Rev 225:91

21. de Mendoza J (1998) Chem Eur J 4:1373
22. Hof F, Rebek J Jr (2002) Proc Nat Acad Sci USA 99:4775
23. Prins LJ, Reinhoudt DN, Timmerman P (2001) Angew Chem Int Ed 40:2382
24. Rebek J Jr (1999) Acc Chem Res 32:278
25. Conn MM, Rebek J Jr (1997) Chem Rev 97:1647
26. Lehn J-M, Dietrich B, Sauvage J-P (1969) Tetrahedron Lett: 2885
27. Dietrich B (1996) In: Vogtle F (ed) Comprehensive supramolecular chemistry, vol 1. Pergamon, p 153
28. Whitlock BJ, Whitlock HW (1990) J Am Chem Soc 112:3910
29. Cram DJ (1983) Science 219:1177
30. Cram DJ, Karbach S, Kim YH, Baczynskyj L, Marti K, Sampson RM, Kalleymeyn GW (1988) J Am Chem Soc 110:2554
31. Chapman RG, Sherman JC (1997) Tetrahedron 53:15,911
32. Cram DJ, Bryant JA, Blanda MT, Vincenti M (1991) J Am Chem Soc 113:2167
33. Cram DJ, Sherman JC, Knobler CB (1991) J Am Chem Soc 113:2194
34. Timmerman P, Verboom W, van Veggel FCJM, van Duynhoven JPM, Reinhoudt DN (1994) Angew Chem Int Ed 33:2345
35. Cram DJ, Tanner ME, Knobler CB (1991) J Am Chem Soc 113:7717
36. Cram DJ, Blanda MT, Paek K, Knobler CB (1992) J Am Chem Soc 114:7765
37. Warmuth R, Yoon J (2001) Acc Chem Res 34:95
38. Cram DJ, Tanner ME, Thomas R (1991) Angew Chem Int Ed 30:1024
39. Warmuth R (1997) Angew Chem Int Ed 36:1347
40. Warmuth R (1998) Chem Commun 59
41. Kurdistani SK, Helgeson RC, Cram DJ (1995) J Am Chem Soc 117:1659
42. Robbins TA, Cram DJ (1993) J Am Chem Soc 115:12,199
43. Paek K, Ihm H, Yun S, Lee HC, No KT (2001) J Org Chem 66:5736
44. Marquez C, Nau WM (2001) Angew Chem Int Ed 40:4387
45. Collet A (1993) Top Curr Chem 165:103
46. Collet A (1996) In: Vogtle F (ed) Comprehensive supramolecular chemistry, vol 2. Pergamon, p 325
47. Bartik K, Luhmer M, Dutasta J-P, Collet A, Reisse J (1998) J Am Chem Soc 120:784
48. McMorran DA, Steel PJ (1998) Angew Chem Int Ed 37:3295
49. Parac TN, Caulder DL, Raymond KN (1998) J Am Chem Soc 120:8003
50. Caulder DL, Powers RE, Parac TN, Raymond KN (1998) Angew Chem Int Ed 37:1840
51. Ziegler M, Brumaghim JL, Raymond KN (2000) Angew Chem Int Ed 39:4119
52. Scherer M, Caulder DL, Johnson DW, Raymond KN (1999) Angew Chem Int Ed 38: 1588
53. Parac TN, Scherer M, Raymond KN (2000) Angew Chem Int Ed 39:1239
54. Su C-Y, Cai Y-P, Chen C-L, Zhang H-X, Kang B-S (2001) J Chem Soc Dalton Trans 359
55. Vilar R, Mingos DMP, White AJP, Williams DJ (1998) Angew Chem Int Ed 37:1258
56. Campos-Fernandez CS, Clerac R, Dunbar KR (1999) Angew Chem Int Ed 38:3477
57. Campos-Fernandez CS, Clerac R, Koomen JM, Russell DH, Dunbar KR (2001) J Am Chem Soc 123:773
58. Fleming JS, Mann KLV, Carraz C-A, Psillakis E, Jeffery JC, McCleverty JA, Ward MD (1998) Angew Chem Int Ed 37:1279
59. Paul RL, Bell ZR, Jeffery JC, McCleverty JA, Ward MD (2002) Proc Nat Acad Sci USA 99:4883
60. Schweiger M, Seidel SR, Arif AM, Stang PJ (2002) Inorg Chem 41:2556
61. Fujita M, Umemoto K, Yoshizawa M, Fujita N, Kusukawa T, Biradha K (2001) Chem Commun 509
62. Kusukawa T, Fujita M (1998) Angew Chem Int Ed 37:3142
63. Fujita M, Oguro D, Miyazawa M, Oka H, Yamaguchi K, Ogura K (1995) Nature 378:469
64. Kusukawa T, Fujita M (1999) J Am Chem Soc 121:1397
65. Kusukawa T, Yoshizawa M, Fujita M (2001) Angew Chem Int Ed 40:1879
66. Yoshizawa M, Kusukawa T, Fujita M, Yamaguchi K (2000) J Am Chem Soc 122:6311

67. Yoshizawa M, Kusukawa T, Fujita M, Sakamoto S, Yamaguchi K (2001) J Am Chem Soc 123:10,454
68. Fujita M, Yoshizawa M, Takeyama Y, Kusukawa T (2002) Angew Chem Int Ed 41:1347
69. MacGillivray LR (2002) Cryst Eng Comm 4:37
70. Aoyagi M, Biradha K, Fujita M (1999) J Am Chem Soc 121:7457
71. Fujita M, Yu S-Y, Kusukawa T, Funaki H, Ogura K, Yamaguchi K (1998) Angew Chem Int Ed 37:2082
72. Yu S-Y, Kusukawa T, Biradha K, Fujita M (2000) J Am Chem Soc 1222:2665
73. Takeda N, Umemoto K, Yamaguchi K, Fujita M (1999) Nature 398:794
74. Umemoto K, Tsukui H, Kusukawa T, Biradha K, Fujita M (2001) Angew Chem Int Ed 40:2620
75. Hiraoka S, Kubota Y, Fujita M (2000) Chem Commun 1509
76. Fujita M, Nagao S, Ogura K (1995) J Am Chem Soc 117:1649
77. Johnson DW, Xu J, Saalfrank RW, Raymond KN (1999) Angew Chem Int Ed 38:2882
78. Liu H-K, Sun W-Y, Ma D-J, Yu K-B, Tang W-X (2000) Chem Commun 591
79. Sun W-Y, Fan J, Okamura T-A, Xie J, Yu K-B, Ueyama N (2001) Chem Eur J 7:2557
80. Olenyuk B, Whiteford JA, Fechtenkotter A, Stang PJ (1999) Nature 398:796
81. Olenyuk B, Levin MD, Whiteford JA, Shield JE, Stang PJ (1999) J Am Chem Soc 121:10,434
82. Kuehl CJ, Kryschenko YK, Radhakrishnan U, Seidel SR, Huang SD, Stang PJ (2002) Proc Nat Acad Sci USA 99:4932
83. Radhakrishnan U, Schweiger M, Stang PJ (2001) Org Lett 3:3141
84. Baxter PNW, Lehn J-M, Baum G, Fenske D (1999) Chem Eur J 5:102
85. Baxter PNW, Lehn J-M, Kneisel BO, Baum G, Fenske D (1999) Chem Eur J 5:113
86. Ikeda A, Ayabe M, Shinkai S, Sakamoto S, Yamaguchi K (2000) Org Lett 2:3707
87. Fox OD, Dalley NK, Harrison RG (1998) J Am Chem Soc 120:7111
88. Zhong Z, Ikeda A, Shinkai S, Sakamoto S, Yamaguchi K (2001) Org Lett 3:1085
89. Fochi F, Jacopozzi P, Wegelius E, Rissanen K, Cozzini P, Marastoni E, Fisicaro E, Manini P, Fokkens R, Dalcanale E (2001) J Am Chem Soc 123:7539
90. Pirondini L, Bertolini F, Cantadori B, Ugozzoli F, Massera C, Dalcanale E (2002) Proc Nat Acad Sci USA 99:4911
91. Park SJ, Hong J-I (2001) Chem Commun 1554
92. Zhong Z, Ikeda A, Ayabe M, Shinkai S, Sakamoto S, Yamaguchi K (2001) J Org Chem 66:1002
93. Ikeda A, Udzu H, Yoshimura M, Shinkai S (2000) Tetrahedron 56:1825
94. Ikeda A, Udzu H, Zhong Z, Shinkai S, Sakamoto S, Yamaguchi K (2001) J Am Chem Soc 123:3872
95. Ikeda A, Yoshimura M, Udzu H, Fukuhara C, Shinkai S (1999) J Am Chem Soc 121:4296
96. Fox OD, Drew MGB, Beer PD (2000) Angew Chem Int Ed 39:136
97. Fox OD, Drew MGB, Wilkinson EJS, Beer PD (2000) Chem Commun 391
98. Beer PD, Berry N, Drew MGB, Fox OD, Padilla-Tosta ME, Patell S (2001) Chem Commun 199
99. Aoki S, Shiro M, Kimura E (2002) Chem Eur J 8:929
100. Drljaca A, Hardie MJ, Ness TJ, Raston CL (2000) Eur J Inorg Chem 2221
101. Hardie MJ, Johnson JA, Raston CL, Webb HR (2000) Chem Commun 849
102. MacGillivray LR, Diamente PR, Reid JL, Ripmeester JA (2000) Chem Commun 359
103. Coppens P, Ma B, Gerlits O, Zhang Y, Kulshrestha P (2002) Cryst Eng Comm 4:302
104. Cave GWV, Hardie MJ, Roberts BA, Raston CL (2001) Eur J Org Chem 3227
105. Kobayashi K, Shirasaka T, Yamaguchi K, Sakamoto S, Horn E, Furukawa N (2000) Chem Commun 41
106. Vreekamp RH, Verboom W, Reinhoudt DN (1996) J Org Chem 61:4282
107. MacGillivray LR, Atwood JL (1997) Nature 389:469
108. Shivanyuk A, Rebek J Jr (2001) Chem Commun 2374
109. Schmuck C (2001) Tetrahedron 57:3063
110. Schmuck C, Heil M (2001) Org Lett 3:1253
111. Atwood JL, Barbour LJ, Jerga A (2001) Chem Commun 2376
112. Rose KN, Barbour LJ, Orr GW, Atwood JL (1998) Chem Commun 407

113. Shivanyuk A, Rissanen K, Kolehmainen E (2000) Chem Commun 1107
114. Shivanyuk A, Paulus EF, Rissanen K, Kolehmainen E, Bohmer V (2001) Chem Eur J 7:1944
115. Murayama K, Aoki K (1998) Chem Commun 607
116. Zeng C-C, Tang Y-L, Zheng Q-Y, Huang L-J, Xin B, Huang Z-T (2001) Tetrahedron Lett 42:6179
117. Grawe T, Schrader T, Gurrath M, Kraft A, Osterod F (2000) J Phys Org Chem 13:670
118. Grawe T, Schrader T, Gurrath M, Kraft A, Osterod F (2000) Org Lett 2:29
119. Grawe T, Schrader T, Zadmard R, Kraft A (2002) J Org Chem 67:3755
120. Zadmard R, Schrader T, Grawe T, Kraft A (2002) Org Lett 4:1687
121. Fiammengo R, Timmerman P, Huskens J, Versluis K, Heck AJR, Reinhoudt DN (2002) Tetrahedron 58:757
122. Corbellini F, Fiammengo R, Timmerman P, Crego-Calama M, Versluis K, Heck AJR, Luyten I, Reinhoudt DN (2002) J Am Chem Soc 124:6569
123. Rebek J Jr (2000) Chem Commun 637
124. Kikuchi Y, Tanaka Y, Sutarto S, Kobayashi K, Toi H, Aoyama Y (1990) J Am Chem Soc 114:10,302
125. Chapman RG, Sherman JC (1998) J Am Chem Soc 120:9818
126. Sherman JC, Chapman RG (1995) J Am Chem Soc 117:9081
127. Chapman RG, Olovsson G, Trotter J, Sherman JC (1998) J Am Chem Soc 120:6252
128. Heinz T, Rudkevich DM, Rebek J Jr (1999) Angew Chem Int Ed 38:1139
129. Heinz T, Rudkevich DM, Rebek J Jr (1998) Nature 394:764
130. Korner SK, Tucci FC, Rudkevich DM, Heinz T, Rebek J Jr (2000) Chem Eur J 6:187
131. Chen J, Rebek J Jr (2002) Org Lett 4:327
132. Hayashida O, Shivanyuk A, Rebek J Jr (2002) Angew Chem Int Ed 41:3423
133. Shivanyuk A, Rebek J Jr (2001) Proc Nat Acad Sci USA 98:7662
134. Shivanyuk A, Rebek J Jr (2001) Chem Commun 2424
135. Atwood JL, Barbour LJ, Jerga A (2002) J Supra Chem Advance Article
136. Kuroda Y, Kawashima A, Hayashi Y, Ogoshi H (1997) J Am Chem Soc 119:4929
137. Turner B, Shterenberg A, Kapon M, Botoshansky M, Suwinska K, Eichen Y (2002) Chem Commun 726
138. Makha M, Hardie MJ, Raston CL (2002) Chem Commun 1446
139. Wyler R, de Mendoza J, Rebek J Jr (1993) Angew Chem Int Ed 32:1699
140. Branda N, Wyler R, Rebek J Jr (1994) Science 263:1267
141. Branda N, Grotzfeld RM, Valdes C, Rebek J Jr (1995) J Am Chem Soc 117:85
142. Pons M, Millet O (2001) Prog Nuc Mag Res Spec 38:267
143. Szabo T, Hilmersson G, Rebek J Jr (1998) J Am Chem Soc 120:6193
144. Kim KM, Park JS, Kim Y-S, Jun YJ, Kang TY, Sohn YS, Jun M-J (2001) Angew Chem Int Ed 40:2458
145. Valdes C, Spitz UP, Toledo LM, Kubik SW, Rebek J Jr (1995) J Am Chem Soc 117:12,733
146. Meissner R, Rebek J Jr, de Mendoza J (1995) Science 270:1485
147. Kang J, Rebek J Jr (1996) Nature 382:239
148. Santamaria J, Martin T, Hilmersson G, Craig SL, Rebek J Jr (1999) Proc Nat Acad Sci USA 96:8344
149. Kang J, Rebek J Jr (1997) Nature 385:50
150. Kang J, Hilmersson G, Santamaria J, Rebek J Jr (1998) J Am Chem Soc 120:3650
151. Kang J, Santamaria J, Hilmersson G, Rebek J Jr (1998) J Am Chem Soc 120:7389
152. Rebek J Jr, Branda N, Grotzfeld RM (1996) Science 271:487
153. Rebek J Jr. http://www.scripps.edu/skaggs/rebek/
154. O'Leary BM, Grotzfeld RM, Rebek J Jr (1997) J Am Chem Soc 119:11,701
155. Lutzen A, Starnes SD, Rudkevich DM, Rebek J Jr (2000) Tett Lett 41:3777
156. Rivera JM, Martin T, Rebek J Jr (2001) J Am Chem Soc 123:5213
157. O'Leary BM, Szabo T, Svenstrup N, Schalley CA, Lutzen A, Schafer M, Rebek J Jr (2001) J Am Chem Soc 123:11,519
158. Anslyn EV, Cabell LA, Best MD, Lavigne JJ, Schneider SE, Perreault DM, Monahan M-K (2001) J Chem Soc Perkin Trans 2:315

159. Steed JW, Abouderbala L, Belcher WJ, Boutelle MG, Cragg PJ, Dhaliwal J, Fabre M, Turner DR, Wallace KJ (2002) Chem Commun 358
160. Hollingsworth MD, Harris KDM (1996) In: Vogtle F (ed) Comprehensive supramolecular chemistry, vol 6. Pergamon, p 177
161. Etter MC, Urbanczyk-Lipkowska Z, Zia-Ebrahimi M, Panunto TW (1990) J Am Chem Soc 112:8415
162. Scheerder J, Fochi M, Engbersen JFJ, Reinhoudt DN (1994) J Org Chem 59:7815
163. Scheerder J, Engbersen JFJ, Casnati A, Ungaro R, Reinhoudt DN (1995) J Org Chem 60:6448
164. Boerrigter H, Grave L, Nissink JWM, Chrisstoffels LAJ, van der Maas JH, Verboom W, de Jong F, Reinhoudt DN (1998) J Org Chem 63:4174
165. Mogck O, Bohmer V, Vogt W (1996) Tetrahedron 52:8489
166. Rebek J Jr, Shimizu KD (1995) Proc Nat Acad Sci USA 92:12,403
167. Scheerder J, Vreekamp RH, Engbersen JFJ, Verboom W, van Duynhoven JPM, Reinhoudt DN (1996) J Org Chem 61:3476
168. Mogck O, Pons M, Bohmer V, Vogt W (1997) J Am Chem Soc 119:5706
169. Jolliffe KA, Calama MC, Fokkens R, Nibbering NMM, Timmerman P, Reinhoudt DN (1998) Angew Chem Int Ed 37:1247
170. Vysotsky MO, Thondorf I, Bohmer V (2000) Angew Chem Int Ed 39:1264
171. Vysotsky MO, Bohmer V (2000) Org Lett 2:3571
172. Cho YL, Rudkevich DM, Rebek J Jr (2000) J Am Chem Soc 122:9868
173. Vysotsky MO, Pop A, Broda F, Thondorf I, Bohmer V (2001) Chem Eur J 7:4403
174. Gonzalez JJ, Ferdani R, Albertini E, Blasco JM, Arduini A, Pochini A, Prados P, de Mendoza J (2000) Chem Eur J 6:73
175. Alajarin M, Lopez-Lazaro A, Pastor A, Prince PD, Steed JW, Arakawa R (2001) Chem Commun 169
176. Alajarin M, Orenes R, Pastor A, Steed JW (2002) J Org Chem 67:7091
177. Rivera JM, Rebek J Jr (2000) J Am Chem Soc 122:7811
178. Brody MS, Schalley CA, Rudkevich DM, Rebek J Jr (1999) Angew Chem Int Ed 38:1640
179. Atwood JL, Szumna A (2002) J Am Chem Soc 124:10,646
180. Atwood JL, Barbour LJ, Jerga A (2002) Proc Nat Acad Sci USA 99:4837

Author Index Volumes 101–108

Alajarin M, see Turner DR (2004) *108*: 97–168
Aldinger F, see Seifert HJ (2002) *101*: 1–58
Aliev AE, Harris KDM (2004) Probing Hydrogen Bonding in Solids Using Solid State NMR Spectroscopy *108*: 1-54
Amstutz N, see Hauser A (2003) *106*: 81–96
Anitha S, Rao KSJ (2003) The Cemplexity of Aluminium-DANN Interactions: Relevance to Alheimer's and Other Neurological Diseases. *104*: 79–98
Aramburu JA, see Moreno M (2003) *106*: 127–152
Atanasov M, Daul CA, Rauzy C (2003) A DFT Based Ligand Field Theory *106*: 97–125
Atwood DA, see Conley B (2003) *104*: 181–193
Atwood DA, Hutchison AR, Zhang Y (2003) Compounds Containing Five-Coordinate Group 13 Elements. *105*: 167-201

Barriuso MT, see Moreno M (2003) *106*: 127–152
Bellandi F, see Contreras RR (2003) *106*: 71–79
Berend K, van der Voet GB, de Wolff FA (2003) Acute Aluminium Intoxication. *104*: 1–58
Bohrer D, see Schetinger MRC (2003) *104*: 99–138
Boulon G (2004) Optical Transitions of Trivalent Neodymium and Chromium Centers in $LiNbO_3$ Crystal Host Material *107*: in preparation
Bowlby BE, Di Bartolo B (2003) Spectroscopy of Trivalent Praseodymium in Barium Yttrium Fluoride *106*: 191–208
Budzelaar PHM, Talarico G (2003) Insertion and β-Hydrogen Transfer at Aluminium. *105*: 141–165
Burrows AD (2004) Crystal Engineering Using Multiple Hydrogen Bonds *108*: 55–96

Cancines P, see Contreras RR (2003) *106*: 71–79
Conley B, Atwood DA (2003) Fluoroaluminate Chemistry. *104*: 181–193
Contreras RR, Suárez T, Reyes M, Bellandi F, Cancines P, Moreno J, Shahgholi M, Di Bilio AJ, Gray HB, Fontal B (2003) Electronic Structures and Reduction Potentials of Cu(II) Complexes of [N,N′-alkyl-bis(ethyl-2-amino-1-cyclopentenecarbothioate)] (alkyl = ethyl, propyl, and butyl) *106*: 71–79

Daul CA, see Atanasov M (2003) *106*: 97–125
Day P (2003) Whereof Man Cannot Speak: Some Scientific Vocabulary of Michael Faraday and Klixbüll Jørgensen *106*: 7–18
Delahaye S, see Hauser A (2003) *106*: 81–96
Di Bartolo B, see Bowlby BE (2003) *106*: 191–208
Di Bilio AJ, see Contreras RR (2003) *106*: 71–79

Fontal B, see Contreras RR (2003) *106*: 71–79
Frenking G, see Lein M (2003) *106*: 181–191
Frühauf S, see Roewer G (2002) *101*: 59–136
Frunzke J, see Lein M (2003) *106*: 181–191

Gray HB, see Contreras RR (2003) *106*: 71–79
Güdel HU, see Wenger OS (2003) *106*: 59–70
Gütlich P (2004) Recent Advances in Spin Crossover Research *107*: in preparation

Harris KDM, see Aliev (2004) *108*: 1–54
Haubner R, Wilhelm M, Weissenbacher R, Lux B (2002) Boron Nitrides - Properties, Synthesis and Applications. *102*: 1–46
Hauser A, Amstutz N, Delahaye S, Sadki A, Schenker S, Sieber R, Zerara M (2003) Fine Tuning the Electronic Properties of $[M(bpy)_3]^{2+}$ Complexes by Chemical Pressure ($M = Fe^{2+}, Ru^{2+}, Co^{2+}$, bpy = 2,2′-bipyridine) *106*: 81–96
Herrmann M, see Petzow G (2002) *102*: 47–166
Herzog U, see Roewer G (2002) *101*: 59–136
Hoggard PE (2003) Angular Overlap Model Parameters *106*: 37–57
Höpfl H (2002) Structure and Bonding in Boron Containing Macrocycles and Cages. *103*: 1–56
Hutchison AR, see Atwood DA (2003) *105*: 167–201

Jansen M, Jäschke B, Jäschke T (2002) Amorphous Multinary Ceramics in the Si-B-N-C System. *101*: 137–192
Jäschke B, see Jansen M (2002) *101*: 137–192
Jäschke T, see Jansen M (2002) *101*: 137–192
Jaworska M, Macyk W, Stasicka Z (2003) Structure, Spectroscopy and Photochemistry of the $[M(\eta^5\text{-}C_5H_5)(CO)_2]_2$ Complexes (M = Fe, Ru) *106*: 153–172

Lein M, Frunzke J, Frenking G (2003) Christian Klixbüll Jørgensen and the Nature of the Chemical Bond in $HArF^t$ *106*: 181–191
Lever B (2004) Ruthenium Complexes of Non-Innocent Ligands; Aspects of Charge Transfer Spectroscopy *107*: in preparation
Linton DJ, Wheatley AEH (2003) The Synthesis and Structural Properties of Aluminium Oxide, Hydroxide and Organooxide Compounds. *105*: 67–139
Lux B, see Haubner R (2002) *102*: 1–46

Macyk W, see Jaworska M (2003) *106*: 153–172
Mahalakshmi L, Stalke D (2002) The R_2M^+ Group 13 Organometallic Fragment Chelated by P-centered Ligands. *103*: 85–116
Moreno J, see Contreras RR (2003) *106*: 71–79
Moreno M, Aramburu JA, Barriuso MT (2003) Electronic Properties and Bonding in Transition Metal Complexes: Influence of Pressure *106*: 127–152
Morita M (2004) Photoluminescence and Excitation Energy Transfer of Rare Earth Ions in Nanoporous Xerogel and Sol-Gel SiO_2 Glasses *107*: in preparation
Morsch VM, see Schetinger MRC (2003) *104*: 99–138
Mossin S, Weihe H (2003) Average One-Center Two-Electron Exchange Integrals and Exchange Interactions *106*: 173–180
Müller E, see Roewer G (2002) *101*: 59–136

Oshiro S (2003) A New Effect of Aluminium on Iron Metabolism in Mammalian Cells *104*: 59–78

Pastor A, see Turner DR (2004) *108*: 97–168
Patočka J, see Strunecká A (2003) *104*: 139–180
Petzow G, Hermann M (2002) Silicon Nitride Ceramics. *102*: 47–166
Power P (2002) Multiple Bonding Between Heavier Group 13 Elements. *103*: 57–84

Rao KSJ, see Anitha S (2003) *104*: 79–98
Rauzy C, see Atanasov (2003) *106*: 97–125
Reber C (2004) Allowed and Forbidden d-d Bands in Octahedral Coordination Compounds: Intensity Borrowing and Interference Dips in Absorption Spectra *107*: in preparation

Reinen D (2004) The Angular Overlap Model and Vibronic Coupling in Treating s-p and d-s Mixing - a DFT Study *107*: in preparation
Reisfeld R (2003) Rare Earth Ions, Their Spectroscopy of Cryptates and Related Complexes in Glasses *106*: 209–237
Reyes M, see Contreras RR (2003) *106*: 71–79
Riesen H (2004) Progress in Hole-Burning Spectroscopy of Coordination Compounds *107*: in preparation
Roewer G, Herzog U, Trommer K, Müller E, Frühauf S (2002) Silicon Carbide - A Survey of Synthetic Approaches, Properties and Applications. *101*: 59–136

Sadki A, see Hauser A (2003) *106*: 81–96
Schäffer CE (2003) Axel Christian Klixbüll Jørgensen (1931-2001) *106*: 1-5
Schäffer CE (2004) Ligand-Field Theory, Reformulated and Revived by Density Functional Theory *107*: in preparation
Schenker S, see Hauser A (2003) *106*: 81–96
Schetinger MRC, Morsch VM, Bohrer D (2003) Aluminium: Interaction with Nucleotides and Nucleotidases and Analytical Aspects of Determination. *104*: 99–138
Schmidtke HH (2003) The Variation of Slater-Condon Parameters F^k and Racah Parameters B and C with Chemical Bonding in Transition Group Complexes *106*: 19–35
Schubert DM (2003) Borates in Industrial Use. *105*: 1–40
Schulz S (2002) Synthesis, Structure and Reactivity of Group 13/15 Compounds Containing the Heavier Elements of Group 15, Sb and Bi *103*: 117–166
Seifert HJ, Aldinger F (2002) Phase Equilibria in the Si-B-C-N System. *101*: 1–58
Shahgholi M, see Contreras RR (2003) *106*: 71–79
Sieber R, see Hauser A (2003) *106*: 81–96
Stalke D, see Mahalakshmi L (2002) *103*: 85–116
Stasicka Z, see Jaworska M (2003) *106*: 153–172
Steed JW, see Turner DR (2004) *108*: 97–168
Strunecká A, Patocka J (2003) Aluminofluoride Complexes in the Etiology of Alzheimer's Disease. *104*: 139–180
Suárez T, see Contreras RR (2003) *106*: 71–79

Talarico G, see Budzelaar PHM (2003) *105*: 141–165
Trommer K, see Roewer G (2002) *101*: 59–136
Turner DR, Pastor A, Alajarin M, Steed JW (2004) Molecular Containers: Design Approaches and Applications *108*: 97–168

Uhl W (2003) Aluminium and Gallium Hydrazides. *105*: 41–66

van der Voet GB, see Berend K (2003) *104*: 1–58

Weihe H, see Mossin S (2003) *106*: 173–180
Weissenbacher R, see Haubner R (2002) *102*: 1–46
Wenger OS, Güdel HU (2003) Influence of Crystal Field Parameters on Near-Infrared to Visible Photon Upconversion in Ti^{2+} and Ni^{2+} Doped Halide Lattices *106*: 59–70
Wheatley AEH, see Linton DJ (2003) *105*: 67–139
Wilhelm M, see Haubner R (2002) *102*: 1–46
de Wolff FA, see Berend K (2003) *104*: 1–58

Zerara M, see Hauser A (2003) *106*: 81–96
Zhang Y, see Atwood DA (2003) *105*: 167–201

Subject Index

Acetanilide 22
N-Acetyl-D,L-valine 27
Acetylene dicarboxylic acid 35
Alanine 13, 42, 43
Alcohols 39
Amides 24, 27, 66, 69
- /triarylphosphine oxides 30
Amidino-*O*-alkylureas 82
Amino acids 12, 15, 42
2-Aminopyridines 70, 75
2-Aminopyrimidine 70
Anion templated synthesis 103, 112
Anion transport 153
Aspartic acid 43

Barbituric acids 83
Benzamide 24
Benzoic acid 33
- –, deuterated 4
- –, dimer 5
Benzoxazine 13
- dimers 32
N-Benzoyl-L-phenylalanine 29
2,2′-Biimidazole 71–73
Bilayer structures 77–79
Bilirubin 32
Biureto 86
o-Bromobenzoic acids 34

C_{60} inclusion 131, 146
Calcium formate, deuterated 7
Calix[4]arene, DNA paired 137
–, multi-component complex 136
–, sulphonato- 133, 139
–, urea substituted 153–162
Calix[6]arene 161
Calix[6]pyrole 146
Cambridge Structural Database 64, 90
Campho[2,3-c]pyrazole 31
Carboxylates 65, 76, 80, 87
Carboxylic acids 33, 59–66, 68, 75
Carboxymethyldethia coenzyme A 29

Carcerands 100
–, guest escape 103
–, solubility 103
–, synthesis 101, 140
Carceroisomerism, in carcerands 104
–, in hemicarcerands 107
Cation-templated synthesis 103, 112
Cavitands 100, 101
Chiral containers 151, 162
o-Chlorobenzoic acids 34
Citrate synthase 29
Crown ethers 62, 88
Cryptands 99
Cryptophanes 108
Cyanuric acid 68, 69, 83, 85, 87
Cyclen 131
Cyclophanes 100

α-Diacetamide 19
Diamines 75
Diaminotriazines 89
Diamondoid structures 60, 62, 67
Dianin's compound, ethanol, clathrate 39
3,5-Dibromo-1*H*-1,2,4-triazole 47
Dicarboxylates 80, 81
3,5-Dichloro-1*H*-1,2,4-triazole 47
Diglycolic acid 34
Diketopiperazines 66
Dimedone (5,5-dimethyl-1,3-cyclo-hexanedione) 20
2,3-Dimethoxybenzoic acid 34
Dimethylmalonic acid 36
3,5-Dimethylpyrazole 44
Diols 75
Disulfonates 79–80
Dithiobiureto 85

Electric field gradient tensors 24
Electrostatic interactions 138, 139
Enzyme catalysis 23
Etter's rules 75
Exchange spectroscopy, two-dimensional 10

Feist's acid 34
Ferrocene-1,1'-diylbis(diphenyl-methanol-d_1) 9
Fumaric acid 26

GDS (guest determining step) 102
Gluconamide 22
Glucopyranosides 129
Glucose 42
Glutamic acid, α/β polymorphs 42
Glycine 43
Glycolurils 67
–, in a 'jelly doughnut' 150
–, in a 'softball' 149, 162
–, in a 'tennis ball' 147
–, triethylbenzene systems 152
–, with phthalocyanines 151
Glycylglycine 27
– monohydrochloride 28
Gramicidin A 13
GS sheets 77–80
Guanidinium 60, 76–80
Guest exchange 104, 111, 122, 125, 132, 149
Guest-templated synthesis 116
– –, anion templated 113–116, 121, 127
– –, cation templated 112, 137
– –, of carcerands 103

Hemicarcerands 104
–, as a catalytic chamber 107
–, expanded portals 106
–, guest polarisability 108
–, metallo-hemicarcerand 130
–, polarisability of cavity 108
–, single portal 104
–, through-shell reactions 107
–, xylene inclusion 106
Heterodimerisation 148, 152, 156–157
Hexabenzocoronene carboxylic acid 36
Hydrates, hydrogen bonded 17
–, water molecules 16
Hydrogen bonding 1, 55
– –, solid state NMR 1
– – geometry, NMR parameters 14
Hydrophobic cavities 108, 117–118, 122, 132
Hydroxybenzaldehydes 20

Imidazoles 44
Imides 68
–, acyclic 19
In-cavity reactions, cycloaddition 141
– –, Diels-Alder 107, 150
– –, lactone interconversions 107
– –, olefin photodimerisation 121
– –, ring inversion 150
Inclusion compounds, urea/thiourea 43
m-Iodobenzoic acid 34
Isomerism, architectural 79
Isonicotinamide 69
Isonicotinic acid 63, 64
Isophthalic acid 60, 61, 64

Janus molecules 58, 59

KOH, solid 17

Leucine 43
Ligands, bifunctional 59, 63, 85, 86
–, 1D-Ligands 110
–, 2D-Ligands 117
Lineshape analysis 8
Liquid crystals 79, 84

Maleic acid 26
Malonic acid 34
Melamine 83–86
Membrane transport 153
Metal phosphates 16
Metal-directed synthesis 109
2-Methylimidazole 47
2-Methylimidazolium cation 47
Minerals 16
Molecular panelling 109, 117
Monolayers 84
Multiple pulse sequence 6

Naphthazarin (5,8-dihydroxy-1,4-naphthoquinone) 31
Nicotinamide 69
Nicotinic acid 63
NMR, ^{2}H NMR lineshape analysis 8
–, ^{2}H NMR spin-alignment techniques 8
NMR parameters, hydrogen bonding geometry 1, 14
Non-covalent synthesis 134
Nucleic acid bases 24

Oligopeptides 30
Orotic acid 86
β-Oxalic acid 18
Oxalic acid dihydrate 15
Oxalurates 87–88
Oxamates 67
Oxamides 68
Oximes 70

Peptides 12, 15, 20, 27, 42
–, gly-containing 17
Phthalate 24

Phthalic acid 13
Piperazinediones 66
pi-pi interactions 69, 70, 84
- -, calix[6]pyrole 146
- -, cyclophanes 100
- -, in panelled structures 121–122
Poly(L-alanine) 25
Polyglycines 25
Polypeptides 12, 15
-, gly-containing 17
Porphyrins 60, 72, 128, 139, 145
Potassium hydrogen malonate 17, 27
Proteins 42
Pseudorotaxanes 62
Pyrazoles 44
-, 3,5-substituted 23
2-Pyridones 67–68
5-(2-Pyridylmethylene)hydantoin 86
Pyrimidinone 87

Resorcinarene, dithiocarbamate complex 131
-, glucopyranoside host 140
-, metal joined 129
-, multi-component complex 134
-, terphenyl complexes 141
Rotation, combined 6

Scaffold structures 109, 110
Selective inversion 11
Self-assembly 55, 109, 133
Self-complementarity 140
Ship-in-a-bottle guests 118–119
Silanols, silica surfaces 12
Solvent containing assemblies 136
Solvent polarity 136
Spin-lattice relaxation 10
- - time measurement 8
Stabilisation, in metal-based cages 112
-, of benzoyl peroxide 141
-, of cyclobutadiene 107
-, of silanol oligomers 119–121
Suberic acid 34
Sulfonates 60, 77–80
Synthons, supramolecular 57, 76

Tectons 58, 59, 84
Terephthalate 80–82
Terephthalic acid 34, 60, 61
Thermodynamic synthesis 109
Thiosemicarbazato 74
Thiosemicarbazide 81–82
Thiourea 43, 67, 81
-, inclusion compounds 43
Thiourea-d_4 44
p-Toluic acid 34
2,4,6-Triaminopyrimidines 83
Triarylphosphine oxides/amides 30
Triazoles 44
Trimesic acid 60, 61, 64
2,4,6-Trimethylpyridine 27
Triphenylmethanol 40
Triphenylsilanol 41
Tris(ureabenzyl)amines 161–162
Tropolone 37

Uracil 84, 85
Urea 24, 43, 64, 67, 68, 76
Urea-substituted calixarene, guest motion 159
- -, in unimolecular capsule 162
- -, inter-/intramolecular bonding 154
- -, kinetic stability 158
- -, mass spectrometry 156
- -, synthesis 153
Ureidopyrimidones 59, 89

Water-formaldehyde 15

The manufacturer's authorised representative in the EU is Springer Nature Customer Service Centre GmbH, Europaplatz 3, 69115 Heidelberg, Germany. If you have any concerns regarding our products, please contact ProductSafety@springernature.com

Printed and bound by CPI Group (UK) Ltd, Croydon, CR0 4YY
01/10/2026
02225951-0005